Forensic Science in Healthcare

Caring for Patients, Preserving the Evidence

Forensic Science in Healthcare

Caring for Patients, Preserving the Evidence

Connie Darnell

CRC Press
Taylor & Francis Group
Boca Raton London New York

CRC Press is an imprint of the
Taylor & Francis Group, an **informa** business

CRC Press
Taylor & Francis Group
6000 Broken Sound Parkway NW, Suite 300
Boca Raton, FL 33487-2742

© 2011 by Taylor and Francis Group, LLC
CRC Press is an imprint of Taylor & Francis Group, an Informa business

No claim to original U.S. Government works

Printed in the United States of America on acid-free paper
10 9 8 7 6 5 4 3 2 1

International Standard Book Number: 978-1-4398-4490-8 (Paperback)

Library of Congress Cataloging-in-Publication Data

Darnell, Connie, author.
    Forensic science in healthcare : caring for patients, preserving the evidence / Connie Darnell.
       p. ; cm.
    Includes bibliographical references and index.
    ISBN 978143984490-8 (pbk. : alkaline paper)
    1. Medical jurisprudence. 2. Victims of crimes--Medical examinations. I. Title.
    [DNLM: 1. Forensic Medicine--methods. 2. Forensic Nursing--methods. W 700]

RA1051.D37 2011
614'.1--dc22   2010044852

Visit the Taylor & Francis Web site at
http://www.taylorandfrancis.com

and the CRC Press Web site at
http://www.crcpress.com

Contents

3   Evidence   105

Preface

This book is written for personnel at all levels within the healthcare spectrum, and provides the basic knowledge and skills for ensuring that a forensic patient's legal rights are protected within the healthcare setting. I have distilled the important concepts and principles applicable to clinical forensic practice, and present them in direct, easy-to-understand language. Following a brief introduction, the text describes how to identify patients with forensic issues, and guides the reader through the practical aspects of evidence preservation, documentation, and legal reporting procedures.

Acknowledgments

My interest in forensics began when I was in college. While taking care of an 18-month-old who had been severely burned by her mother, I was appalled to learn that the child would be returned to her mother "in consideration of the *mother's* mental health." Finding there were few laws to protect children, I vowed to advocate for the rights of the vulnerable and abused. After more than 10 years, it would be impossible to acknowledge all those who have helped make this book a reality. There are several people, however, who stand out. First, that young child who motivated me to begin this journey—it is really for her, and all others like her, that this book was written.

I would like to thank Janet Barber for her support, guidance, and friendship and for making this book become a reality, and her husband, Don Duval, for his quiet support from the background. I thank Dr. Patrick Besant-Matthews for his willingness to share his expertise, enthusiasm, uncommon knowledge, photographs, and friendship over the years and for teaching me much of what I know. I also thank Dr. William Smock for permission to use some of his photographs, and for his never-ending support of those who follow in his footsteps. I also thank Robert Witter for his input and guidance on electrical burns and hazards.

I especially thank Zech Robinson who has unselfishly given his time and expertise with computers and who kept me sane when I became overwhelmed by electronic demons, Jackie Kerwin and the Silverton Public Library, and George Romero, who completely redid our electrical system when ice avalanching off the roof wiped it out at a critical point in preparing this manuscript. I also acknowledge Bruce Elliot, who has always been my knight in shining armor, and last but certainly not least, my husband for his kindness and support all of my adult life. To them and to all those who listened, critiqued, consulted, and in general supported me in this endeavor, I will forever be indebted.

Author

 **Connie Darnell** is a charter member of the International Association of Forensic Nurses and has been involved in forensics since the early 1990s. She has worked as a deputy field death investigator for the New Mexico Office of the Medical Investigator and is a trained sexual assault nurse examiner. She has taught introductory undergraduate and continuing education forensic nursing classes at the University of New Mexico and has given numerous presentations to healthcare groups, local law enforcement, and volunteer fire departments.

She lives with her husband at their homes in rural New Mexico and Alaska, where she is a member of the local volunteer fire and rescue departments. She is currently employed part time in Santa Fe, New Mexico.

Introduction to Forensics

1

I thought forensics was for law enforcement and death investigators. What does it have to do with me?

Increased popularity of *forensic* * programs has led to increased interest in the forensic sciences. Many believe forensics to be limited to matters involving death.† This is simply not true. In fact, most forensic cases involve people who are alive, thus leading to the term *living forensics*.

Clinical forensics is the application of forensic science to medical situations. A patient exhibiting symptoms of carbon monoxide poisoning is an example. Is there a product liability issue? Is this an accident, negligence, or an intentional attempt to harm or kill? Another example might be a pregnant woman who is exposed to radiation. Both of these examples require a detailed investigation into the circumstances surrounding the incident as well as determination of the health consequences to the individual.

Motor vehicle accidents, sexual assault, and other interpersonal violence are commonplace. Substance abuse is widespread and crosses all socioeconomic boundaries. Terrorism and mass casualty incidents are a part of our daily topics of concern. All of these have forensic issues and implications for healthcare providers.

Violence and traumatic incidents in America pose a mental health, public health, and public safety dilemma. Once limited to assessing and caring for a patient's physical or psychological needs, healthcare's responsibility now encompasses the identification and treatment of those impacted by crime, violence, and other intentional or unintentional trauma. The inclusion of criminal and civil liability into healthcare has made forensic knowledge and practice a legal imperative. In the past, healthcare's failure to recognize victims of crime and violence was a significant factor in reduced reporting to authorities. Our justice system now requires healthcare providers to think and act with full regard for their ethical and legal responsibilities to the patient

---

* The word forensic comes from the Latin forensic, meaning "of a market or forum: public." The term evolved into its current usage, of "belonging to courts of judicature or to public discussion and debate." Thus, forensic medicine (also known as medical jurisprudence) is a science dealing with the relationship and application of medical facts to legal problems.

† As contrasted with forensic pathology, living forensics is the application of forensic science to surviving victims.

and to the justice system. It is the healthcare provider's observations, coupled with an intuitive sense of the intangible, which may bring forensic issues to light. From initial assessment to discharge planning, forensic issues must be considered and, if applicable, included in the delivery of patient care.

Collaborating with others along the continuum of care is a routine occurrence for providers of healthcare. Primary providers are uniquely positioned to function as a liaison between the medical community and law enforcement, the judicial system, insurance companies, social service organizations, community agencies, lawmakers, bureaucrats, and others. The creation of multidisciplinary teams has proven to be the most effective way to manage the effects of trauma and violence. Not only does a team approach provide for a better patient outcome, it helps ensure the safety of the healthcare provider. The very nature of the healthcare profession lends itself to follow through into the legal arena. Future victims benefit from the variety of viewpoints presented. Resources are maximized, expenses are minimized, and efficiency in the delivery of services is increased.

To maintain credibility as professionals, it is essential that professional providers of care redefine their role as *patient advocate*. Providers at all levels must learn how to advocate for patients medically, psychologically, and forensically by being unbiased seekers of the truth. It is imperative that the healthcare provider not make value judgments or take sides. Rather, we must be impartial observers and recorders of the facts.

Red Flags

In addition to the obvious (motor vehicle accident [MVA], gunshot wound, etc.) there are several "red flags" that may indicate that a patient has forensic issues or is the victim of interpersonal violence. Among these are:

- Unexplained or unwitnessed traumatic injuries or illness—the patient and/or accompanying individuals have no explanation at all or say no one saw what happened
- Unusual or avoidable delay in seeking medical care
- Multiple injuries—different kinds of injury or multiple sites are involved
- Multiple injuries indicating more than one incident—wounds of differing ages or stages of healing, or multiple presentations for healthcare for the same or similar injuries or illnesses
- Patterned injuries—injuries suggesting a weapon or instrument was used

- Injuries that are inconsistent with the account given or that are unlikely to have been caused as the scenario has been told—they simply do not "fit," or the patient and accompanying individuals give differing accounts of what happened
- Any time there is suspicion that the injury was not accidentally caused—a gut feeling that something just isn't right

The Evolution of Crime, Violence, and Crime Detection

In medieval England, the sheriff was charged with protecting the interests of the king. He was concerned with crimes committed against secular or religious authority and not necessarily crimes perpetrated against common people. Today the sheriff's job is to encourage people to obey the law, apprehend those who do not, and investigate the circumstances.

The office of the coroner was formalized in England in the twelfth century. As clinical forensic medicine matured, the concept of the *police surgeon* came into being. In 1842, the London city police department extended the role of law enforcement to include crime scene investigation. The creation of detective investigators was a significant improvement because it gave detectives the power to interview witnesses, collect and preserve evidence, and work collaboratively with other law enforcement professionals and the court system.

Living Forensics

Dr. Henry McNamara, a New York medical examiner, believed that, in addition to medical needs, survivors of catastrophic events have legal issues. He also believed that healthcare had an important and undeveloped role to play. He introduced the concept of *living forensics*. Within a few short years, the first clinical forensic medicine training program was begun in Louisville, Kentucky. The International Association of Forensic Nursing (IAFN) was founded in 1992 and shortly thereafter formally recognized by the American Academy of Forensic Sciences (AAFS) and the American Nurses' Association (ANA). Standards of forensic nursing practice were soon adopted and published (McNamara, 1986).

In 1994, the Violence against Women Act (VAWA) was passed by Congress. Several hundred sexual assault nurse examiner (SANE) programs are now in place in the United States and numerous programs have been started in other countries. Employment for nurses as death investigators, clinical forensic nurse specialists, forensic correctional/psychiatric nurses, and legal nurse consultants is now a reality. Thousands of former nurses are now practicing law as nurse attorneys. Pediatric and geriatric nurses, emergency medical technicians (EMTs), and first responders now possess basic forensic skills.

The Scope of Crime and Violence in Modern Society*

The Crime Index (FBI, 2002) estimated that there were 1.6 million violent crimes reported to law enforcement in 2001 (USDOJ, FBI, OJP, OVC, BJS, NCVS, cited in NVAA, 2002). According to the FBI, one violent crime occurred in the United States every 19 seconds, one forcible rape occurred every 5 minutes, and one murder occurred every 29 minutes.

Tragically, less than half of violent crimes and only about one-third of all crimes are ever reported to the police. Healthcare's failure to recognize victims and the impact on individual lives has been a factor in reduced incidence of reporting to authorities. Our failure to understand this has led to tragic and lifelong consequences for victims and their friends and families.

A disaster is defined as an event or situation causing ruin or failure (Compact Oxford English Dictionary, 2005). Using this definition, the current level of violence in America is clearly a public health and safety disaster. This tragedy is a preventable circumstance that not only affects individual victims, friends, and families, but also touches communities and the country as a whole. The events of September 11, 2001, are a dramatic testament to the short- and long-term impact such events can make.

In 1998, there were 8.1 million crimes of violence in the United States, only 46% of which were ever reported to law enforcement. In 1994, 1.4 million people were treated in emergency departments for suspected or confirmed interpersonal violence (BJS, August, 1997), 1.6 million violent crimes were reported to law enforcement (FBI, 1998) and an estimated 3.7 million adult women were victims of some type of sexual or aggravated assault during a one-year period (Tjaden and Thoennes, The National Women's Study: Research in brief, 1992).[†]

In our fast-paced lives, people succumb to increasing pressures of time and money. Substance abuse is widespread and has a multitude of significant consequences. These consequences are not limited to the health implications for the user or addict. Families and the stranger with whom the substance abuser comes in contact are also affected. Drugs and alcohol are implicated in motor vehicle accidents, interpersonal violence, sexual assault, homicide,

---

* Begun in 1995, the National Victim Assistance Academy is a week-long, university-based training course on victimology, victims' rights and victim services. (NVAA, 2002).
† All of the above statistics are reported in the 2002 NVAA Manual.

suicide, and other injurious behaviors. Fetal alcohol syndrome and the crack-addicted newborns are just two of the many adverse consequences visited upon the innocent.

Violence influences how people view the world. Trust is the glue that binds human relationships. Violence makes individuals wary of unfamiliar people. The result is an erosion of one's sense of personal safety, leading to skepticism, cynicism, and self-imposed isolation. The overall quality of life is decreased because human interaction is stifled.

Since the terrorist acts of 9/11, violence—or, more importantly, the *fear* of violence—has changed the way Americans think and live. People no longer feel safe at night, even in their own neighborhoods—indeed, in their own homes. Many have installed some type of home security system or purchased a weapon for self-protection. People no longer feel safe when traveling. Fear of violent encounters restricts individuals' inclination to go where they wish, when they wish, how they wish. The once hated inconveniences of heightened security are now accepted. Our view of humanity has undergone a significant change.

The Financial Cost of Violence

Key Point:
Forensic cases have financial and emotional costs as well as medical ones.

In 1996, the National Institute of Justice released a comprehensive report on the cost of violent crime (NVAA Manual, 1999; US Department of Justice [USDOJ], FBI, OJP, OVC). Data was gathered from criminal justice agencies, medical professionals, crime victim compensation programs, and crime victims themselves. The financial impact of violence amounts to well over $400 billion each year (National Institute of Justice [NIJ] 1997–1998 Academy Text Supplement Ch. 1 p. 3). As much as 20% of mental health expenditures in this country can be attributed to treatment of victims alone* (National Institute of Justice [NIJ] Cost of Victimization, February 1996). The U.S. government pays nearly $20 billion in health insurance payments and for services to victims. Crime costs private insurers approximately $45 billion annually (NIJ, 1996).

* This estimate does not include figures for mental health treatment of offenders.

The Emotional Cost of Violence

Key Point:
Psychological injuries impact more than just the victim: friends, family, coworkers, and others can also be significantly affected.

Victims may have psychological wounds equal to or greater than the physical ones. The professional provider must be sensitive to those emotional needs. Understanding a victim's reaction to crisis enables the healthcare provider to provide appropriate psychological care. It is important to remember that, to those impacted by trauma and violence, this is not just a case, it is a tragedy and their lives will never be the same. Even though every crime is not necessarily violent, a sense of violation remains. Survivors—whether they are direct victims or those affected by extension—may exhibit signs of posttraumatic stress disorder (PTSD), not only at the traumatic moment, but days, weeks, or months after the events have occurred.

The Role of Healthcare

Survivors of catastrophic events encompass a wide variety of scenarios, including interpersonal violence, sexual assault, unnatural death, legal and custodial disputes, legislation, malpractice, workplace injury, drug and alcohol dependence, terrorist incidents, and others. Living forensics has implications for healthcare workers in many settings, including hospitals, clinics, insurance companies, government agencies, law enforcement offices, attorneys' offices, judicial settings, industrial settings, legislative offices, and literally on the street.

Violence, trauma, and other liability-related situations are pervasive in modern culture. It is imperative that healthcare providers develop a "forensic antenna" as part of their basic repertoire. Are the injuries consistent with the explanation given? All healthcare providers must learn the basic elements of forensic content, including how to document forensically, how to collect and preserve evidence, reporting requirements, and how to defend or explain medical situations in court.

Key Point:
Personal security concerns related to both the patient and the healthcare staff must be addressed throughout all phases of the forensic case management.

Living forensic patients typically have experienced serious traumatic events, and need to be confident regarding their personal safety. Once a sense of personal security has been established, the patient, families, and friends need to be reassured that they are not being judged. Central to managing the short- and long-term effects of traumatic events is to ease feelings of guilt for the situation they find themselves in. When patients feel someone is listening to them in a nonjudgmental way, fear and anxiety are reduced. Patients begin to believe their needs will be satisfied. Explaining roles, procedures, and treatments provides reassurance and engenders trust. Rather than probative questioning, it is better to allow the patient to relate their experience in their own terms.

The presence of family and friends is important to victims, and those needs must also be considered. Whether the patient is victim, perpetrator, or family, the healthcare provider must be courteous and respectful. These individuals need support, but boundaries must be set on behavior. Disruptive behavior can be minimized by a calm, professional, caring, and patient-centered approach. Out-of-control individuals may have to be restrained or removed from a particular clinical setting. This is *not* the responsibility of the healthcare provider, but rather the responsibility of security or law enforcement.

Today's professional healthcare provider understands basic anatomy and physiology and can help identify a mechanism of injury or cause of death. Nurses' expertise in normal growth and development, the disease processes, medical terminology, and recognition of mental, emotional, and physical disabilities, helps interpret aspects of medical care to the legal community. Assessing a patient's condition and documenting the care given become second nature to the experienced healthcare provider.

Nurses and other healthcare professionals also possess communication skills that enable them to help individuals through a crisis, and to care for patients in uncontrolled and unpredictable environments. They use their knowledge of human psychology on a daily basis and are skillful at interacting with grieving individuals and noncompliant patients.

Key Point:
Nurses function in an ever-expanding variety of roles. Collaboration with other healthcare providers and agencies is important in the continuum of care.

The frontline practitioner is in a unique position to collect and preserve evidence. Frontline practitioners may be the ones who initiate the all-important *chain of custody*. An example occurred during a forensic nursing class I taught. The class was condensed into two sessions, one month apart. In the interim between sessions, one of my students was working in the emergency department when a woman arrived with a "self-inflicted" gunshot wound. Based on the information this nurse had received in the first class, she was suspicious enough to package the woman's clothing appropriately, initiate the chain of custody, and call the police. Their initial response was, "Where did you learn to do *this*?" Since the chain of custody had been properly initiated, police took the sealed, packaged evidence to the evidence locker, and, upon inspection, arrested the woman's boyfriend. She died on the operating table before she was able to talk to law enforcement; he was charged and ultimately convicted of murder.

Nurses are educators. Nurses routinely teach patients and their families how to manage their health, prevent disease, and ward off complications. Forensic patients provide the nurse with a unique opportunity to educate other medical professionals in the recognition and treatment of forensic issues as well as inform patients about things such as risky behavior.

Legal nurse consultants review and analyze medical records for the legal community, insurance companies, and others. In certain instances, legal nurse consultants may also testify as expert witnesses. Nurses often serve as consultants for private and governmental agencies who draft rules and legislation. They are ideally positioned to help prevent trauma and injury through research, education, and outreach. In Minnesota, a nurse ran for the office of coroner and won. Once in office, she noticed the rate of teen suicide was higher than expected. She collaborated with local schools to set up suicide prevention programs. The result was a significant reduction in the number of teen suicides. Recently, in South Carolina, nurse coroners noticed an increasing number of deaths in the elderly. Utilizing creative, grassroots techniques, those nurses began an active campaign to educate the community and find ways to better support its elders.

Terrorism has caused fear and apprehension on a scale previously unknown in America. Unsafe products, environmental, occupational, and epidemiological hazards create scenarios that change lives. It is the front-line healthcare provider who first recognizes a cluster of similar symptoms or illnesses that might be related. Recently, a number of individuals from a single apartment complex presented to emergency rooms with symptoms of carbon monoxide poisoning. Utilizing their "forensic antenna" (increased suspiciousness factor), healthcare providers collaborated with others, leading to the discovery of faulty heaters and use of ovens to keep warm in cold weather. The city forced the owner of the apartments to fix the problem or be shut down. The city also provided hotel vouchers to tenants so they had a warm, safe environment until the problem was rectified. The owner was ultimately charged with multiple crimes.

Key Point:
All healthcare workers have a responsibility to develop and maintain a forensic antenna in order to identify possible legal issues associated with their patients, including violations of human rights.

Standards of Care

In order for a healthcare discipline to be credible, it must provide an objective and measurable foundation upon which quality of care, performance evaluation, and peer review can be achieved. In order to ensure quality patient care, healthcare providers are held to standards set by their peers. Standards are the foundation upon which patient care is provided. The *standard of care* relates to the direct and indirect care patients receive; the *standard of professional performance* relates to professional practice. Adherence to accepted standards generates respect, promotes further growth of the profession through research, and encourages collaboration with other professions.

The document that outlines and describes the standards for forensic nursing practice was published by the International Association of Forensic Nurses in 2009 (IAFN, 2009). This organization has also released a similar document specifically related to intimate partner violence. These professional performance standards require that nurses are educated in a prescribed curriculum and attain basic competencies in legal as well as clinical nursing issues. They also describe the collaborative interfaces with other professional disciplines, and serve to define certain professional boundaries. The quality and effectiveness of forensic practice must be systematically and objectively evaluated in order to ensure quality patient care and the evolution of forensics as part of healthcare delivery.

The Nursing Process

Nursing care is dependent upon a logical scientific process that structures care practices from the initial patient contact through a final evaluation when the nurse–patient relationship is severed. This basic framework is called the *nursing process*, and its components can be applied to victims of trauma, violence, and crime. Most related disciplines, such as physicians or emergency medical providers, use a similar approach.

Key Point:
Steps of the nursing process can be directly applied to forensic issues in patient care.

Assessment: Data Collection

The first step in the process is the collection of data. Data is used to assess the individual's physical and psychological status and to formulate a nursing diagnosis. Forensic nursing diagnoses incorporate the legal aspects of a patient's situation into the standard nursing diagnoses. Once nursing diagnoses have been made, the forensic nurse identifies expected outcomes and implements a plan of action. The plan is finally evaluated and modified, starting the process all over again.

Using the process of elimination, where does the assessment process lead? Has the possibility of forensic issues been identified in the problem list or have they been ruled out? If so, does the plan of care include forensic concerns?

Initial assessment includes the interview, physical exam, review of records, observations, and collaboration with staff and other healthcare providers. Once this data has been collected, it is assembled into a logical thought pattern so conclusions can be drawn from the signs, symptoms, and inferences (clues) gathered. The healthcare provider's experience and intuition are critical in interpreting the factual data. Interpretation and evaluation are the next step in developing nursing diagnoses and forming of a plan of care.

Forensic patients require balancing of competing interests. Examples of this conflict are the delivery of direct care versus the requirements of law, or patient confidentiality versus society's interests. Because of the traumatic and often urgent nature of forensic patients' conditions, medical assessment and intervention must often be done quickly. Legal consequences, however, require rapid and thorough forensic assessment and documentation.

Nursing Diagnosis

The North America Nursing Diagnosis Association (NANDA) has approved an official definition of nursing diagnosis as "a clinical judgment about individual, family, or community responses to actual or potential health problems/life processes" (Carpenito, 1993). Some approved nursing diagnoses include potential for violence, potential for injury, pain related to injuries, altered health maintenance, ineffective individual/family coping, and so forth.

Planning

Patient planning is the method by which nursing diagnoses can be addressed. Planning basics dictate that life threats be resolved first. Other issues are secondary and should be prioritized in a logical order based on patient safety and need. Short- and long-term goals need to be established and specific measurable objectives identified in order to achieve those goals. Objectives are simply descriptions of what is expected to happen to or for the patient after nursing intervention has been implemented. Strategies are problem-solving methods utilized to achieve the objectives. Again, these may be short-term methods, such as providing a stuffed animal to a child to reduce stress, or long-term methods, such as referring the family to counseling as a part of the discharge plan.

Implementation

Implementation includes the specific tasks needed to accomplish the objectives. The nursing process requires that the patient, family, and/or others be informed and/or consulted. Delivery of care must provide for patient safety, be outcome oriented, and be consistent with stated objectives. Once the care plan has been implemented, patient and family responses need to be documented.

Evaluation

Evaluation revolves around whether the patient and family are able to meet the specific goals and objectives. Will the patient be safe in the immediate future? Have mechanisms been put in place so the patient can cope with what has or is happening? Is there a mechanism in place to evaluate the care given, referrals made, and so on? Long term, has the patient or family changed in a positive way as a result of nursing intervention? How can that be evaluated? By whom? If family dynamics have not been changed, what safeguards have been put in place to help ensure the future safety and security of the patient, children, pets, or other members of this family?

Summary

Healthcare's responsibility was once limited to assessing and caring for a patient's physical or psychological needs. The identification of injuries and deaths that may have medical-legal ramifications is now one of our responsibilities. Nurses all too often depended on the patients to tell them why they sought treatment or on the doctor to make a medical diagnosis. All healthcare providers must now use their scientific and intuitive skills to assess and intervene on behalf of today's forensic patients. Mastery of forensic content and skills provides a framework to build an intuitive forensic antenna—the art of critical "noticing" that enhances one's insight and refines the suspiciousness factor (Winfrey and Smith, 1999). Appendix 1.1 lists general signs of abuse, neglect, or exploitation.

The healthcare provider does not work in a vacuum. Communicating, collaborating, networking, and sharing of resources, information, research technology, and expertise are all methods we can use to achieve the common goal of improving the health and safety of individuals and society as a whole. The creation of multidisciplinary teams has proven to be the most effective way to manage the effects of trauma and violence. A team approach provides for better patient outcomes, and allows a variety of viewpoints to be presented and enhances the safety of providers. Additionally, resources are maximized by contributing to the reduction of expenses while increasing efficiency in the delivery of services.

Healthcare personnel do not make judgments about forensic issues, nor do they take sides of either the victims or those who have offended. They care for those who are subject to ongoing abuse, neglect, or violence and bring to light those invisible victims who would have remained vulnerable without intervention. Well-educated and skilled individuals within healthcare serve as a vital link between meeting human needs and facilitating justice.

Appendix 1.1: Signs of Abuse, Neglect, and Exploitation*

General:
    History inconsistent with physical findings
    Delay between injury and treatment
    Multiple healthcare providers
    Over-attentive partner, parent, or caregiver
    Suspicious death or suicide (including attempts)

Physical and Sexual Abuse:
    Fractures and trauma injuries (including teeth and mouth)
    Bilateral, multiple, or patterned injuries
    Injuries in various stages of healing
    Inappropriate sexual behavior of a child
    Sleep disturbances
    Injuries to central or sexual parts of the body
    Inappropriate use of physical or chemical restraints

Psychological Abuse:
    Extremely passive, apologetic
    Isolated or restricted contact with others
    Acting out, aggressive or regressive behavior
    Fear of parent, partner, child, or caregiver
    Demeaning language, tone of voice, disrespect from parent, partner,
        child, or caregiver
    Child witness to domestic violence

Neglect/Self-Neglect:
    Poor hygiene
    Inappropriately or inadequately dressed
    Hungry, malnourished, and/or dehydrated
    Medications withheld or improperly administered
    Hypochondria
    Depressed, withdrawn, and/or apathetic
    Destructive behavior toward self or others

Exploitation:
    Sexual abuse or assault
    Victim of financial abuse or fraud
    Mismanagement of resources by caregiver or other person

* Courtesy Deborah Williams, Clinical Forensic Nurse Specialist, Louisville, Kentucky.

References

Carpenito, L. 1993. p. 5. In Murry, P. and Stein, M. Child Abuse, 4th ed., Clinical Nursing Series, p. 186. South Easton, MA: Western Schools Press. (1997).

Child Maltreatment. 2007. (Washington, DC: U.S. Department of Health and Human Services, Administration on Children, Youth and Families, 2009), 23–25. http://www.acf.hhs.gov/prorams/cb/pubs/cm07/cm07.pdf (accessed September 2, 2010).

Ching-Tung Wang and John Holton. 2007. "Total Estimated Cost of Child Abuse and Neglect in the United States," Washington, DC: Prevent Child Abuse America, 4, 5.

Coleman, G., M. Gaboury, M. Murray, and A. Seymour. 1999. *National Victim Assistance Academy Manual 1999*. Grant No. 95-MU-GX-K002(S-4). USDOJ, OJP, OVC, NVAA 1999. 149.101.22.150/OVC/assist/nvaa99/welcome/html (accessed August 28, 2010)

Compact Oxford English Dictionary of Current English, 3d Edition. 2005. Oxford, UK: Oxford University Press.

Cost of Victimization. 1996. *National Institute of Justice Journal*, February 1996 http://www. ncjrs.gov/odffukes/nijj 230.pdf (accessed September 2, 2010).

Finkelhor, David. 2009. "Violence, Abuse, and Crime Exposure in a National Sample of Children and Youth," *Pediatrics* 124, no. 5. http://ovc.ncjrs.gov/ncvrw2010/pdf/6_StatisticalOverviews.pdf (accessed September 2, 2010).

IAFN. 1997. *Scope and Standards of Forensic Nursing Practice*. International Association of Forensic Nurses (IAFN)/American Nurses Association (ANA). Washington, DC: American Nurses Publishing.

Kilpatrick, D. G., C. N. Edmunds, and A. K. Seymour. 1992. *Rape in America: A report to the nation*. Arlington, VA: National Center for Victims of Crime. Charleston, SC: Medical University of South Carolina.

McNamara, Henry. *Living Forensics*. (seminar pamphlet), Ulster County, NY. Office of the Medical Examiner, 1986. Cited in Lynch, Virginia A., Clinical Forensic Nursing: A new perspective in the management of crime victims from trauma to trial. *Critical Care Nursing Clinics of North America*, 7(3), 489-507, p. 491. (September 1995).

National Association of Crime Victim Compensation Boards, "Crime Victim Compensation Helps Victims," Alexandria, VA; NACVCB, 2009, http://www.nacvcb.org (accessed September 2, 2010).

National Institute of Justice Journal. Cost of Victimization. February 1996. Accessed September 9, 2010 at http://www.ncjs.gov/pdffiles/hijj_230.pdf.

National Victim Assistance Academy (NVAA) Manual. 2002. Seminars held simultaneously at 1) California State University, Fresno, CA; 2) Medical University of South Carolina, Charleston, SC; and 3) Washburn University, Topeka, KS. http://www.ncjrs.gov/ovc_archives/nvaa2002/aboutbook.html (accessed September 9, 2010). Individual chapters can be accessed at http://www.NVAAchapter/html/2002/chapter1_1.html.

NMS Labs. 2010. http://www.nmslab.com (accessed September 9, 2010).

Rand, Michael. "Criminal Victimization, 2008," Washington, DC: Bureau of Justice Statistics, (2009), 1, http://www.ojp.usdoj.gov/bjs/pub/pdf/cv08.pdf (accessed September 2, 2010).

Tjaden, P., and Thoennes, N. 1998. *The National Women's Study: Research in Brief.* Office on Women's Health, Department of Health and Human Services. Content last updated March 1, 2009.

U.S. Department of Justice (USDOJ), Federal Bureau of Investigation (FBI), Office of Justice Programs (OJP), Office of Victims of Crime (OVC), *National Victim Assistance Academy Manual (NVAA), 1999,* Chapter 1, Scope of Crime, Historical Review of the Victims' Rights Discipline. (June 1999). http://www.ojp.usdoj.gov/ovc/assist/nvaa1999/chapter1 (accessed January 18,2003.)

U.S. Department of Justice (USDOJ), Federal Bureau of Investigation (FBI), Office of Justice Programs (OJP), Office of Victims of Crime (OVC), 1997-1998 Academy Text Supplement. http://www.ojp.usdoj.gov/ovc/assist/nvaa/supp/a-ch1.htm (accessed September 2, 2010). Citing Miller, T., Cohen, J., and Wiersema, B. February 1996. Victim Costs and Consequences: A new look. Washington, DC: National Institute of Justice, U.S. Department of Justice.

U.S. Department of Justice (USDOJ), Federal Bureau of Investigation (FBI), Office of Justice Programs (OJP), Office of Victims of Crime (OVC). (2002). *National Victim's Assistance Academy (NVAA), 2002.* http://www.ojp.usdoj.gov/assist/nvaa2002/chaapter1

U.S. Department of Justice, (USDOJ). 2002. Bureau of Justice Statistics. National Crime Victimization Survey (NCVS), July 2002. Washington, DC.

Walker, S. 1999. Scope of Crime: Historical review of the victim's rights discipline. *National Victim Assistance Academy Manual.*

Wikipedia. http://en.wikipedia.org/wiki/Disaster (accessed August 22, 2010).

Winfrey, M. E. and A. R. Smith, A. 1999. The suspiciousness factor: Critical care nursing and forensics. *Critical Care Nursing Quarterly*, 22(1), 1–7. http://www.preventchildabuse.org/about_us/media_releases/pcaa_pew_economic_impact_study(final.pdf (accessed September 2, 2010).

Documentation

2

Medical Documentation

Before the Joint Commission on the Accreditation of Healthcare Organizations (JCAHO or "Joint Commission") was established, nursing documentation was haphazard and not considered to be an important part of the medical record. Often, nurse's notes were destroyed once the patient was discharged. The Joint Commission promoted formation of nursing standards and required that nurse's notes be a permanent part of the patient's record. This milestone improved overall patient care in several ways.

- It provided a 24-hour snapshot of the patient's day chronicling treatments given and changes in the patient's condition.
- It became an excellent source of information in investigating possible criminal and civil activity.
- It provided a time sequence of events and documented the involvement of caregivers.
- It tied other elements in the chart together, giving them context and meaning.

Creation of nursing diagnoses in the mid-1970s allowed nurses to define patient problems and act on them. Unlike medical diagnoses, nursing diagnoses emphasize a patient's biophysical, psychological, psychosocial, environmental, cultural, learning, and discharge planning needs. Nursing diagnoses identify strengths, weaknesses, and potential problems as the patient progresses toward wellness. Forensic issues lend themselves well to the use of nursing diagnoses in preparing a plan of patient care.

It is important both medically and forensically for healthcare providers to use appropriate language and terminology, correct grammar, spelling, punctuation, and logical organization. Development of standard forms and checklists have made charting easier, faster, and more efficient for nurses, but in the rush to be efficient, charting can become routine, critical thinking is less common, and errors of omission are easier to make.

Nursing documentation must accurately reflect a patient's physical and emotional status, the care provided, and the patient's response to that care. Nurses' observations, assessments, and plan of care must be recorded. Nurses' charting must reflect the care given, the quality of that care, and evidence of each healthcare provider's participation in, and accountability for, providing that care. Nursing documentation must also include aspects of evidence collection and preservation. Forensically speaking, when medicolegal issues are identified, nurses must be acutely aware that everything they do is reflected in documentation, and may be scrutinized in a court of law. That scrutiny often occurs at a time far enough in the future that memories have faded and whatever is in the written record is all that is left of the moment.

When charting the collection of evidence, the nurse must include a descriptive listing of

- Item or items collected
- Date and time of collection
- Name and title of person or persons collecting the evidence
- Location of evidence collected
- Substantiating sketches, body diagrams, and/or photographs
- Item or case number so the evidence can be cross-referenced later on

If an interpreter was used, the name, qualifications, and language must be included. Information included must be legible, clearly and concisely stated, objective, and accurate. Handling of evidentiary items is discussed in further detail in Chapter 3.

Recognition and documentation of injuries is an important step in the examination and treatment of victims of trauma and crime. The following discussion of the different types of injury will be helpful in understanding how to approach the narrative and body diagram portions of the nursing record. Important details include:

- Clear identification of location, using correct anatomical terms
- Specifying on which side of the body the injury is found
- Using terms such as abrasion, laceration, contusion, burn, stab wound, and so forth, accurately and appropriately
- Recording the size (length and width) of each injury
- Describing an approximate shape of each wound (circular, ovoid, linear, triangular, patterned, irregularly shaped, etc.)
- Using simple, commonly understood terms to describe the color of wounds
- Back up the narrative with sketches, diagrams, and photographs

Wounds are often difficult to describe, particularly for the novice or inexperienced nurse. Body diagrams can be very helpful adjuncts to the narrative if done properly (see Figure 2.1). Body diagrams and photographs are an adjunct to, not a substitution for, the nurse's narrative. They are, however, an

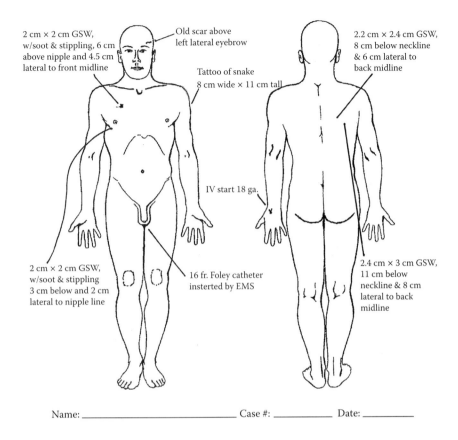

2 cm × 2 cm GSW, w/soot & stippling, 6 cm above nipple and 4.5 cm lateral to front midline

Old scar above left lateral eyebrow

Tattoo of snake 8 cm wide × 11 cm tall

2.2 cm × 2.4 cm GSW, 8 cm below neckline & 6 cm lateral to back midline

IV start 18 ga.

2 cm × 2 cm GSW, w/soot & stippling 3 cm below and 2 cm lateral to nipple line

16 fr. Foley catheter inserted by EMS

2.4 cm × 3 cm GSW, 11 cm below neckline & 8 cm lateral to back midline

Name: _____ Case #: _____ Date: _____

Figure 2.1 Body drawing with annotations.

excellent way to visualize the distribution, pattern, and/or overall picture as seen through the nurse's eyes. Diagrams and sketches need not be complex, but should convey the location and type of all injuries in as much detail as possible. Samples are included as Appendices 2.1 through 2.13. All medical treatments (e.g., attempted IV sites, EKG lead placement, etc.) should also be noted. Use of standard symbols is acceptable if a key is included on the body diagram sheet. Regardless of whether injuries are seen only on one body surface, both anterior and posterior views should be included. The *absence* of wounds may be as important later as their presence. In addition to body diagrams, additional sketches are acceptable if needed to accurately and completely illustrate the nature and extent of wounds to the body.

General Physical Examination

The first step is to look at the victim or patient to evaluate his or her general overall state. Also note the patient's behavior and reactions.

Key Point:
Psychosocial needs should be included in the secondary survey.

Treatment of the patient's immediate physical needs must be addressed before or simultaneously with gathering the history. However, whether the patient's injuries are the result of accident or intentional harm, these patients have been traumatized. If possible, obtain the medical and forensic history in a quiet and private setting. Always be cognizant of the patient's psychological needs prior to and during information gathering. Remember the patient's body may be a crime scene. As with all patients, take and document initial vital signs and perform a quick initial assessment. Note and record the patient's general appearance, affect, and attire, using objective terms.

The healthcare provider's job is to be an observer of the facts and to record those observations as accurately and completely as possible. Avoid use of judgmental terminology. Credibility of documentation is diminished when an individual's characteristics, mental status, behavior, or any events are described in a critical or biased manner.

Carefully document signs and symptoms, allergies, medications, and pertinent medical history. Do not forget to note the patient's height, weight, hair, and eye and skin color. Ask straightforward questions regarding location, date and time of event, previous forensic history, nature of the incident/assault, activities since the incident/assault, and any information known about the perpetrator.

Carefully remove and package all clothing as described in the evidence collection chapter, noting their general condition and any evidence of trauma.

Conduct a complete visual examination, starting at the head and working toward the feet, observing the body for signs of trauma. Use an ultraviolet (Wood's Lamp) or alternate light source to help identify subtle injuries such as bite marks, rope marks or burns, or recent contusions. Observe for signs of redness, swelling, tenderness, or indurations. In cases of sexual assault, record hygiene and bathing activities immediately prior to, during, or after the assault.

Collect all dried and moist secretions. Note the method used for collection and disposition of any specimens collected.

If there are no relevant physical findings, physiological changes, or evidence of foreign material, their absence should be clearly stated in the record.

Wound Documentation

Wounds provide a great deal of information about the mechanism of injury and information for treatment. They provide law enforcement with physical evidence for prosecution.

Key Point:
The essence of good surface documentation is:

1. Where it is
2. How big it is
3. What it looks like
4. The angle of entry (if appropriate)

Measure as accurately as possible, using standard anatomical landmarks (e.g., left corner of mouth, 3 inches below right nipple, etc.). An American Board of Forensic Odontology (ABFO) rule is the standard, especially for photography, and is shown in Figure 2.2.

It is important that all body surfaces are examined, including the genital area. Have a second healthcare professional in attendance when conducting the physical examination. The purpose is threefold: (1) to verify the examiner's findings, (2) to reduce embarrassment and to provide a sense of security for the victim, and (3) to protect the examiner from future false or misleading statements by the patient.

In addition to looking for wounds—cuts, bruises, lacerations, and abrasions—you're going to note the *absence* of signs that, based on the patient's story, one would expect to see. Foremost in a forensic nurse's mind is the question: Are the injuries consistent with the story? Does what you see correspond to the explanation being given?

Clothing is an important source of information regarding the mechanism of injury. Clothing will be discussed in greater detail in the chapter on evidence.

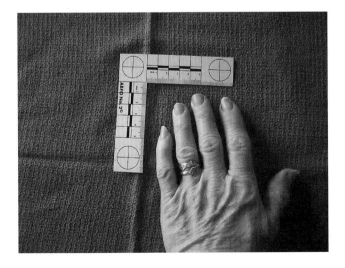

Figure 2.2 Hand with ABFO rule.

Figure 2.3 Using a quarter to approximate wound size.

Except in life-threatening situations, observation and documentation of physical injuries must be made prior to medical intervention. Much is lost in the process of cleaning, examining, medical or surgical intervention, and the effects of the passage of time (the healing process or infection). Initial observations include location, size, and appearance/character. Document each wound with respect to fixed landmarks such as ears, elbows, belly button, and so on. When documenting size, use of a ruler or scale is preferred; for example, "irregular abrasion measuring 2 cm × 3.5 cm." However, comparison with a common object (nickel, quarter, pencil lead, etc.) can be used. See Figure 2.3. When documenting appearance, use words that denote texture, color, whether the wound was wet or dry, bloody, dirty, and so on (e.g., "grains of yellow sand imbedded in abrasion").

Physical injuries fall into one of four basic categories: blunt force trauma, sharp force trauma, gunshot or fast force wounds, and thermal injuries.

Blunt Force Injuries

Blunt force injuries fall into one or more of four basic categories. They may be seen singly or in combination with other blunt force trauma.

- *Abrasions*—scratches, grazes, and bites
- *Contusions*—bruises
- *Lacerations*—tears
- *Fractures of bone*—complete or incomplete disruption of bone integrity

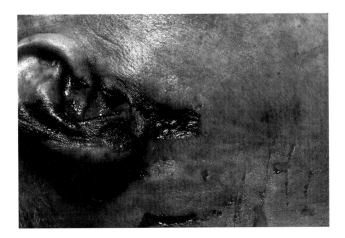

Figure 2.4 Post-mortem illustration of pattern injury. Victim was struck with water pipe. Note the eight vertical thread markings on the right side of the image. (Four incised wounds below thread markings were incurred during pre-autopsy shaving procedure required to fully expose the pattern injury associated with the homicide.) Courtesy Dr. Patrick Besant-Matthews. Used with permission.

Abrasions

Abrasions are the removal of the outermost layer of skin. Two forces work simultaneously to create abrasions: *compression* or downward pressure on the skin, and *sliding,* the longitudinal force along the surface of the skin. Because direction can be determined, this information may be important.

In addition to indicating contact with a rough surface, abrasions also signify the *exact* point of contact. In the living patient, abrasions will crust over (scab) and darken. See Figure 2.4. After death, they will dry and darken due to the lack of circulation. This darkening can lead to the false interpretation that the abrasion resulted from burning or bruising.

Scratches are long narrow abrasions, like those made by a fingernail, thorn, or cat's claw (see Figure 2.5). *Grazes* are wider (see Figure 2.6). An example of a graze is the large abrasion on the knee when someone falls on the ground.

The form and appearance of abrasions should be noted: (1) to help determine the mechanism and circumstances of injury, and (2) because they may be of future significance in judicial or liability determinations. In a sliding abrasion known as *shelving,* tissues on the surface may be pushed toward one end, like dirt pushed by a bulldozer, indicating the direction of force (see Figure 2.7).

Abrasions may also exhibit a characteristic pattern, indicating contact with a specific object, such as a rug. This type of injury is called a *pattern injury.* Pattern injuries are blunt force injuries where the instrument leaves an impression that reflects the characteristics of the instrument used (see Figures 2.8 through 2.20).

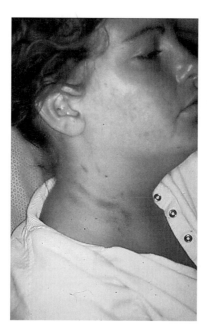

Figure 2.5 Scratch on woman's neck. Courtesy Dr. William S. Smock. Used with permission.

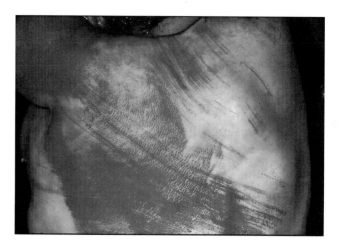

Figure 2.6 Grazes. Courtesy Dr. Patrick Besant-Matthews. Used with permission. (See color insert following p. 202.)

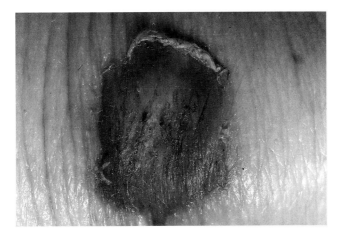

Figure 2.7 Abrasion with accumulated tissue at one end showing direction of force. Courtesy Dr. Patrick Besant-Matthews. Used with permission. (See color insert following p. 202.)

Abrasions may be seen in conjunction with bruises and lacerations. If the forces are sufficient to produce an abrasion, they may also distort the underlying tissue enough to tear small vessels and other tissue.

Key Point:
An abrasion indicates the exact site of contact or impact between skin and a rough surface or object, and may exhibit a characteristic pattern of the object contacted.

Key Point:
If sliding forces were involved, the direction of force may be determined by the accumulation of tissue at one edge of the abrasion.

Contusions

Contusions (bruises)* result from leakage (large or small) of blood from vessels into the surrounding tissue (see Figure 2.21. The application of enough blunt force to distort the soft tissues will tear one or more vessels, resulting in leakage of blood into the tissues. Usually, these tears occur in capillaries, but if one or more of the vessels are big enough, the leak may be sufficient to cause swelling (see Figure 2.22). In cases such as strangulation, pinpoint

---

* Ecchymosis is the leakage of blood into skin or mucous membranes and is seen in some disease states or in the elderly. It is the result of a physical condition and not the result of trauma. Contusions are exclusively the result of a traumatic event (however small). The term contusion is not synonymous with the term ecchymosis.

Figure 2.8 Pattern abrasion from blow with pipe. Courtesy Dr. Patrick Besant-Matthews. Used with permission.

Figure 2.9 Threaded pipe used to strike victim. Courtesy Dr. Patrick Besant-Matthews. Used with permission.

bruising, known as *petechiae*, can be seen in delicate areas such as the eyelids (see Figure 2.23).

Key Point:
Only an abrasion or patterned injury in or near the bruising itself will indicate the exact point of contact.

Bruises may not become visible for hours, or even days, due to shock or delayed escape of fluids from a blood vessel. Because a contusion requires leakage of blood from a vessel, significant blunt force does not necessarily

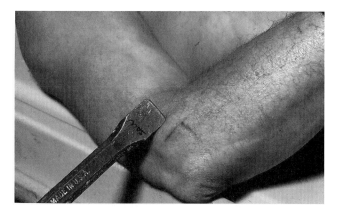

Figure 2.10 Pattern injury to right elbow; possibly defensive wound. Courtesy Dr. Patrick Besant-Matthews. Used with permission.

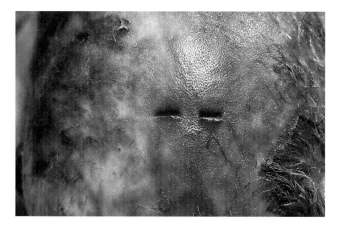

Figure 2.11 Injury from fork. Courtesy Dr. Patrick Besant-Matthews. Used with permission.

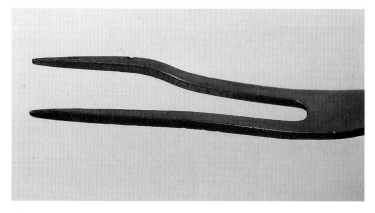

Figure 2.12 Fork used to inflict blunt-force wound seen in Figure 2.11. Courtesy Dr. Patrick Besant-Matthews. Used with permission.

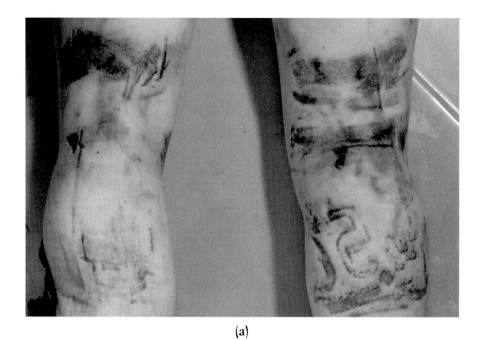

(a)

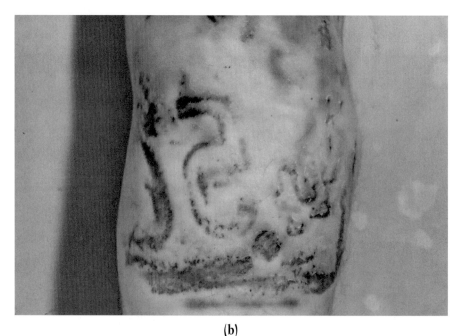

(b)

Figure 2.13a,b Pattern injury. Imprint of license plate on victim's legs where struck by vehicle. Courtesy Dr. Patrick Besant-Matthews. Used with permission. (See color insert following p. 202.)

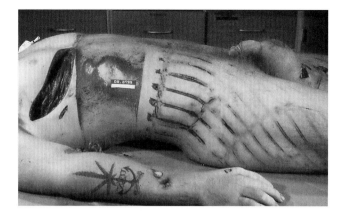

Figure 2.14 Pattern injury from grate. Courtesy Dr. Patrick Besant-Matthews. Used with permission.

result in the formation of a bruise. Discoloration may not necessarily appear at the place where force was applied because blood may have to track around muscles, fascia layers, or other structures on its route to the surface where it can be seen. An abrasion near a contusion will indicate the point at which forces were applied (the exact point of contact), but a contusion itself is not an indicator of the point of contact. It is important for the healthcare provider to understand that the intensity and/or duration of force are difficult to estimate unless abrasions, lacerations, or other features are present.

If a vessel laceration (tear) is significant, blood will tend to escape through the open wound rather than into the surrounding tissues.

A fresh bruise usually appears as a reddish area (the color of oxygenated blood), but like vessels, it may appear blue. Coloring of bruises depends on depth, amount of fat, lighting, and of course, age. As bruises age, they turn

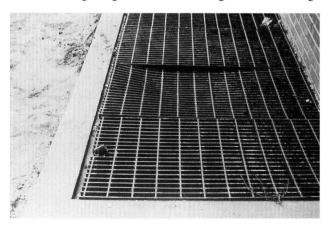

Figure 2.15 Grate associated with wounds seen in Figure 2.14. Courtesy Dr. Patrick Besant-Matthews. Used with permission.

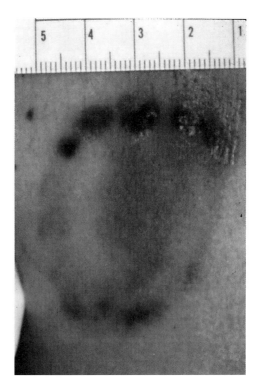

Figure 2.16 Human bite mark. File Photo. (From Catanese, C. (ed.) *Color Atlas of Forensic Medicine and Pathology*, Taylor & Francis/CRC Press, Boca Raton, FL, 2009, p. 207. With permission.)

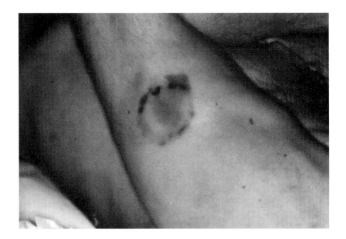

Figure 2.17 Human bite mark. File photo. (From Catanese, C. (ed.) *Color Atlas of Forensic Medicine and Pathology*, Taylor & Francis/CRC Press, Boca Raton, FL, 2009, p. 207. With permission.) (See color insert following p. 202.)

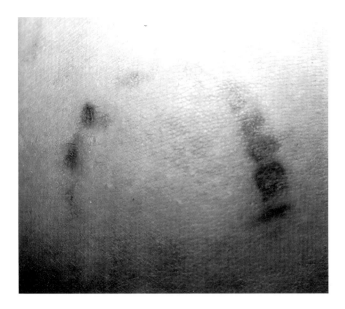

Figure 2.18 Human bite mark. File photo. (From Catanese, C. (ed.) *Color Atlas of Forensic Medicine and Pathology*, Taylor & Francis/CRC Press, Boca Raton, FL, 2009, p. 207. With permission.)

purplish. Finally, as the blood pigments break down, the sequence of colors passes through those of a ripening banana—green to yellow to brown—until the discoloration fades. The rate at which a bruise resolves depends on many factors, including the quantity of blood originally released, the effectiveness of the local circulation, the location on the body, and the individual's

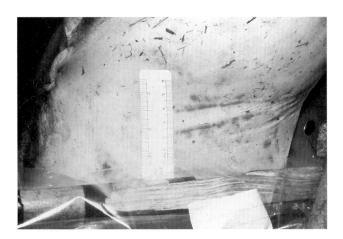

Figure 2.19 Pattern injury from cue stick at bottom of photo. Courtesy Dr. Patrick Besant-Matthews. Used with permission. (See color insert following p. 202.)

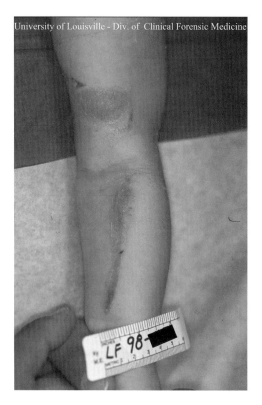

Figure 2.20 Pattern injury. Burn from curling iron. Note: Arm was folded over curling iron. Courtesy Dr. William Smock. Used with permission. (See color insert following p. 202.)

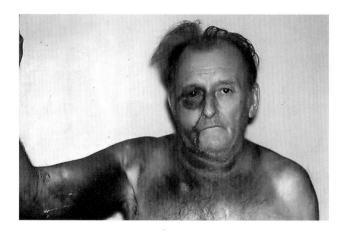

Figure 2.21 Beaten man with multiple bruises. Courtesy Janet Barber. Used with permission.

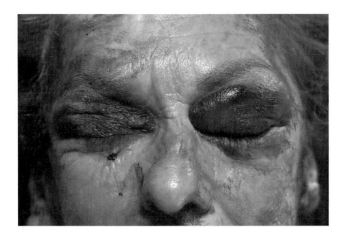

Figure 2.22 Raccoon eyes. Courtesy of Dr. Patrick Besant-Matthews. Used with permission.

age, general physical activity, and condition. Therefore, *estimating the age of bruises is very difficult and should be avoided.*

Documentation of bruising should be limited to size, shape, color, and location. Multiple bruises, especially those of varying colors, are an important fact and should be clearly documented. Bruises of different ages may be adjacent to, or overlap, one another. If the injury is the result of a single event, ask yourself why bruises of different colors are seen simultaneously.

Bruising can also be masked by skin coloration and may be almost invisible in the heavily tanned or naturally dark-skinned individual. Bruises are easily overlooked in areas where blood has been forced, or has settled, or in

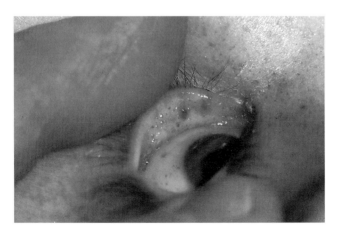

Figure 2.23 Petechiae. Courtesy Janet Barber. Used with permission. (See color insert following p. 202.)

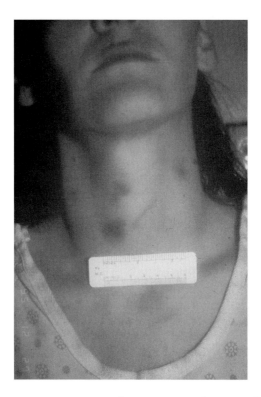

Figure 2.24 Fingertip contusions from attempted strangulation. Courtesy Dr. William Smock. Used with permission. (See color insert following p. 202.)

areas in which circulation is failing. It is generally easier for blood to escape into loose tissues and fat; therefore, bruising is more common in certain parts of the body, after weight loss, in obesity, at the extremes of age, and is more probable if there is any disease of the blood vessels themselves or the patient is on medication that affects clotting. Simply put, some individuals bruise more easily than others.

Bruises, by their very nature, may not become visible immediately. Important internal injuries may exhibit little or no external sign of bruising. The appearance of bruising depends on blood accumulating near enough to the skin to be seen. In some cases, this requires that the blood move from the point where it leaked out of the vessel to its final destination. For this reason, reexamination of a victim a day or two after initial injury may be valuable.

Distribution of bruises is significant. Small bruises around the neck or on a limb may be the only external physical signs of violence (see Figure 2.24). This is especially true in cases of domestic violence and sexual assault.

Postmortem bruising is possible if a body is roughly handled or dropped very shortly after death. For this to occur, however, enough blood must still be present in the tissues and free to move, and gravity must be able to act.

Bruising is accentuated in the presence of any bleeding disease (hemophilia, leukemia, scurvy) and in those taking medications, especially aspirin or other drugs with anticoagulant properties. Some antidepressants such as Zoloft inhibit blood platelets and bruising in unusual locations may be seen. The appearance of blotchy skin occurs with some skin diseases and in those individuals who use prednisone. Sometimes these skin conditions resemble bruising.

Medical treatments, such as advanced cardiac life support, respirators, transfusions, antibiotics, and some drugs, can also alter the formation of bruises. A thorough assessment is necessary to determine if any of these factors are present.

Lacerations

The term *laceration* is one of the most commonly misused terms in modern medicine. Soft tissue injuries are either tears or cuts. Lacerations (tears) are the result of *blunt* forces; for example, a tear on a boxer's forehead is the result of blunt forces (the other boxer's glove) and is a laceration, not a cut. Cuts (incisions) are the result of *sharp* forces.* As with bruising/ecchymosis, the terms are not interchangeable.

Tissue deformities resulting from blunt force trauma include tearing, ripping, crushing, overstretching, pulling apart, bending, and shearing. They are not the result of sharp forces.

Key Point:
Lacerations have ragged, irregular margins and may contain foreign material; cuts have clean margins and rarely contain foreign material. If the wound is linear, close inspection and good history taking will help determine whether the wound was caused by blunt or sharp forces.

Skin lacerations are frequently found over bony prominences. This is because the skin is relatively fixed and less able to move when stressed (see Figure 2.25). Inside the body, organs and arteries such as the spleen, liver, and aorta are most frequently lacerated (in this case, torn) at points of relative immobility.

It can be difficult to determine whether a wound is the result of blunt or sharp forces, or a combination. A skin laceration typically has an irregular margin. Look closely at Figure 2.26. Although the wound appears to have clean edges, there are skin tags all along the edges, making this a laceration. The margins around lacerations often display evidence of scraping and bruising. Because tissues are torn apart, the separation is frequently incomplete.

---

* Description of a superficial wound to the face is dependent on the causative instrument: It is classified as a laceration if the forces are blunt (a hand, fork, or other object) and is classified as a cut if the forces are sharp (a knife or razor).

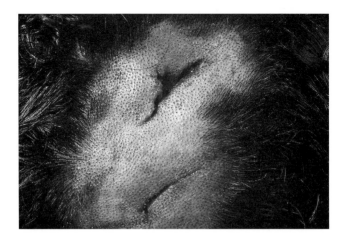

Figure 2.25 Scalp laceration. Courtesy Dr. Patrick Besant-Matthews. Used with permission.

Stronger tissue elements, such as little blood vessels, nerves, and connective tissue strands, survive to bridge or span the gap from one part or side of the wound to the other (see Figures 2.27 and 2.28). When you see bridging, think "blunt force injury."

Bridging is most easily seen at the corners or deep within a wound. A blunt injury made with a linear object may initially appear to be a straight line, but upon closer inspection, the edges are irregular and stronger tissue components "bridge" the edges of the wound.

Forensically, lacerations give clues as to the kind of surface that impacted the skin. Because they are the result of blunt impact, lacerations are clinically important in two ways. First, these wounds are more susceptible to infection

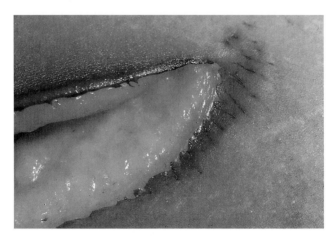

Figure 2.26 Laceration over soft tissue. Courtesy Dr. Patrick Besant-Matthews. Used with permission. (See color insert following p. 202.)

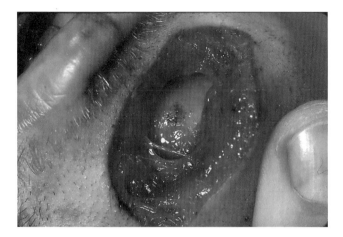

Figure 2.27 Bridging. Courtesy Dr. Patrick Besant-Matthews. Used with permission. (See color insert following p. 202.)

because the crushing forces involved result in less blood flow to the area. Second, they are more vulnerable to infection because they are more likely to be contaminated with foreign material such as dirt, glass, fibers, and paint chips. These foreign materials may be of evidentiary value as trace evidence or foreign objects.

Sharp Force Injuries

Sharp force injuries fall into two basic categories: *cuts* or *incisions* and *stab wounds*.

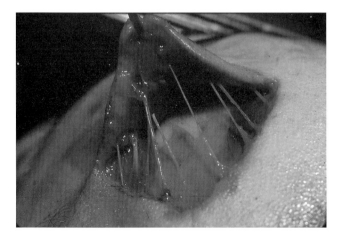

Figure 2.28 Bridging. Courtesy Dr. Patrick Besant-Matthews. Used with permission.

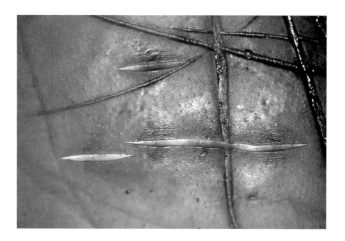

Figure 2.29 Cuts on abdomen. Courtesy Dr. Patrick Besant-Matthews. Used with permission. (See color insert following p. 202.)

Cuts or Incisions

A *cut* is sustained when a sharp object makes contact with the skin with sufficient pressure to divide it. The force necessary to produce a cut depends on (1) the sharpness of the cutting instrument and (2) the resistance of the skin and overlying objects (such as clothing).

Incisions, cuts, and incised wounds are linear and generally longer than they are deep.* The force that creates a cut has enough inward pressure to create the cut, but the primary force is longitudinal along the surface of the skin. Cuts may be deeper at one end, suggesting left or right handedness. Cuts have clean edges and there is no bridging of tissue (see Figures 2.29, 2.30, and 2.31).

Because they tend to bleed freely, blood escapes rather than collecting under the skin. Consequently, little or no bruising may be visible. If vessels are completely severed, they may be able to retract, and may not bleed much at all. If the object is sufficiently sharp, hair and other small structures will also be cut.

Trace evidence is rarely found in a cut.

Stab Wounds

Stab wounds result whenever a sufficiently sharp and narrow object is forced inward. Stab wounds are deeper than they are long (see Figure 2.32). There may be some longitudinal motion, but the primary force is inward. The force

---

* Although cuts are generally linear, they may be irregular if the skin was creased, wrinkled, or affected by clothing.

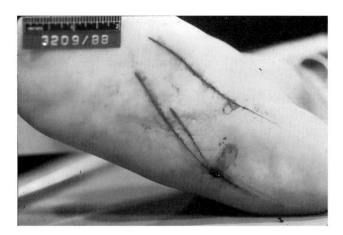

Figure 2.30 Linear cuts with puncture wound to bottom of foot. Courtesy Dr. Patrick Besant-Matthews. Used with permission.

may be a thrust or it may be due to falling on or being impaled by an object capable of penetration. Again, the two critical elements are force and sharpness of the object.

The skin itself offers most of the resistance to a foreign object. Once it has been penetrated, the amount of force required to continue travel is diminished, unless tissues such as cartilage or bone are encountered.

If an object is not sufficiently sharp or if it tapers and becomes thicker as it enters, the skin may stretch, resulting in an abrasion at the wound margin. Stab wounds are primarily sharp in character. It is possible, however, to force a relatively blunt object into the body. In such an instance, the wound is best classified as a laceration (see Figure 2.33).

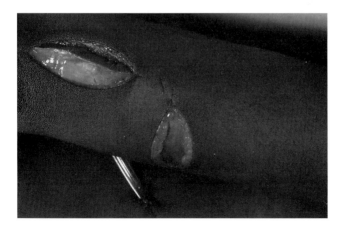

Figure 2.31 Cuts to arm. Probably defensive wounds. Courtesy Dr. Patrick Besant-Matthews. Used with permission. (See color insert following p. 202.)

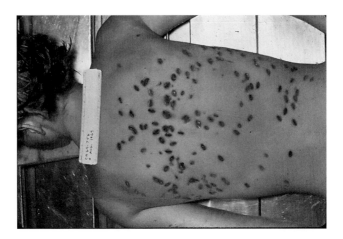

Figure 2.32 Multiple stab wounds to back. Courtesy Dr. Patrick Besant-Matthews. Used with permission.

Stab wounds are more likely to reflect the causative instrument or weapon than are cuts. For example, a fairly thick knife blade with only one sharp edge will tend to leave a defect that has a cleanly cut acute angle at one end and a more squared-off or slightly torn, angular appearance at the other (see Figure 2.34).

The exact depth of penetration cannot be determined unless the stabbing object is available. Areas of abrasion near the point of entry indicate that whatever weapon was used went in as far as was possible. Such stab wounds have a high potential for deep injury. In soft areas such as the abdomen, the skin is flexible enough to allow a stabbing object to reach much deeper than

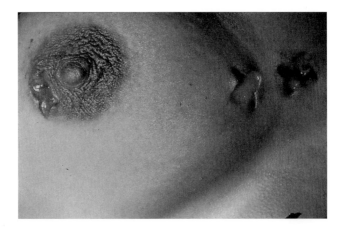

Figure 2.33 Stab wound from Phillips screwdriver-type instrument. Courtesy Dr. Patrick Besant-Matthews. Used with permission.

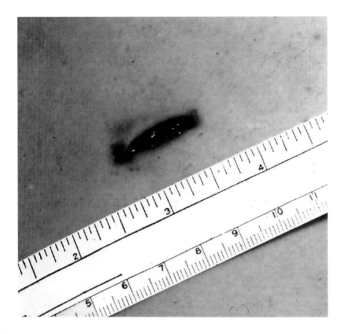

Figure 2.34 Single-edge stab wound. (From DiMaio, V.J.M., and Dana, S.E., *Wounds Handbook of Forensic Pathology, Second Edition*, p. 110. Boca Raton, FL: Taylor & Francis, 2006.)

the length of the object itself. Always look for evidence that a knife or tool was forced in as far as its handle.

Stabbing instruments may occasionally break off inside a wound, especially if bone is encountered* (see Figure 2.35). Recovery and retention of such broken pieces should be done and the pieces handled as potential evidence. They may be vital to effective prosecution of a criminal case, or to prove the cause of an accidental injury.

Important aspects of sharp wounds to remember are:

- Cuts are generally wider than they are deep, and stab wounds are generally deeper than they are wide.
- Injury to vital internal structures may not be immediately apparent. Internal bleeding may result in precipitous collapse of the patient.
- Stab wounds may reflect the causative instrument or weapon.
- There is relatively little abrasion of wound margins unless the weapon is tapered and it wedges/stretches the skin on its way in.
- Generally, there is no bridging in cuts or stab wounds.

---

* Modern knives are more flexible and stronger than those made in some other countries, so this phenomenon is relatively rare in the United States.

(a)

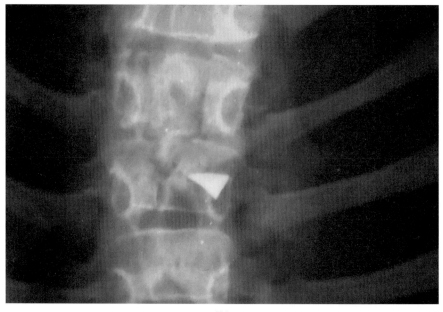

(b)

Figure 2.35 a) Knife with broken tip. b) Broken tip in x-ray. Courtesy Dr. Patrick Besant-Matthews. Used with permission.

Document the surface dimension, depth, and direction of these wounds if at all possible. Beveling or shelving of wound margins is a clue to the internal direction of a wound track.

In the event that a chest tube or other device is placed through a suitably positioned injury,* it is important that this be clearly indicated in the record. Otherwise, it may be difficult to properly interpret the wounds inflicted by an assailant. This is precisely what happened in the shooting of President Kennedy: a tracheotomy tube was placed into his neck wound. Because no exit was identified, misinterpretation and speculation about the number of bullets that hit the president exists to this day.

Mixed Blunt and Sharp Injuries

Victims of assault, motor vehicle accidents, and other multiple traumas may suffer both blunt and sharp injuries. For instance, a person involved in a motor vehicle accident may be cut by a piece of broken glass, impaled by a gear-shift lever (blunt injury that looks like a stab wound), and receive multiple blunt traumas to the chest from the steering wheel and/or dashboard.

Fast Force Injuries (Gunshot Wounds)

Laceration from a fast force projectile is the crushing or tearing of tissue (blunt trauma) as the result of forces related to the transfer of energy from the penetration of a moving object (bullet or pellets; see Figure 2.36). Bullets of greater mass fired at high velocity have the greatest potential for causing

Figure 2.36 Bullets with lands and grooves. Courtesy Dr. Patrick Besant-Matthews. Used with permission.

---

* An exit wound in President John F. Kennedy's neck was used to insert an endotracheal tube, thus causing confusion as to the number of shots fired.

injury. Projectiles from handguns have relatively low velocity (800–1,500 feet/second). Bullets from rifles have up to three times the velocity of those from handguns. Shotguns produce low- to medium-velocity projectiles, but do more damage at close range because of the increased number of projectiles and larger total volume. When assessing the victim of a gunshot, it is important to remember that there is rarely damage to a single structure.

For example, a young man was shot in the abdomen with a .22 caliber rifle. Although no vital structures were hit, the shockwave created significant cavitation of surrounding tissues. This cavitation caused a large amount of tissue damage and subsequent massive rupture of cellular membranes. The patient developed irreversible lactic acid acidosis and died several days later.

A great deal of information can be gleaned from observing gunshot wounds, including the range of fire, angle of entry, and type of projectile(s), and any intervening object between the firearm and victim's body.

When examining a gunshot wound, any evidence should be photographed before the wound is cleaned or disturbed by medical treatment. Parallel narrative documentation as well as a body diagram, should be incorporated into the medical record.

When projectiles are secured as evidence, it is critical that they not be marked in any way. Projectiles should be handled the same as any other evidence, not adding or subtracting from the original condition. If it is necessary to use forceps when removing projectiles, the tips must be covered with portions of a rubber catheter or other soft material that will not mark the projectile (see Figure 2.37).

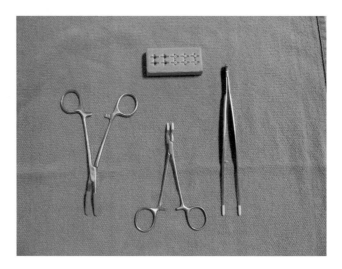

Figure 2.37 Forceps and hemostats with soft tip covers. (See color insert following p. 202.)

Regardless of the type of firearm used, there are three things that need to be evaluated and documented: range of fire, angle of entry, evidence or observations regarding the type of projectile and any intervening materials that may have been between the weapon and the victim.

Range of Fire

The range of fire relates to the distance between the end of the barrel or muzzle of the firearm and the target. We know how far burned and unburned grains of gunpowder travel, so range of fire can be determined (see Figure 2.38).

Contact range gunshot wounds occur when the muzzle of a gun is touching the target at the instant it is fired. In these situations, the powder residue is deposited in the wound track itself (see Figure 2.39). If the muzzle is pressed tightly against the target, all the powder residue is deposited in the wound track; if the muzzle is not tightly sealed against the target, some of the powder and gas escapes into the surrounding surface (see Figure 2.40). In both cases, the entrance wound is modified by the gas and powder residue forced into it. The degree of modification is primarily a function of the amount of gas entering the tissues—the more gas, the more stretching, tearing, and so on. The degree of wound modification is also determined by the underlying anatomy. Blasting of gas into tissues that lie over a hard surface (such as the

Figure 2.38 Gun firing, illustrating soot and unburned grains of powder. Courtesy Dr. Patrick Besant-Matthews. Used with permission.

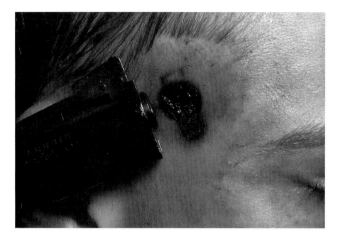

Figure 2.39 Contact gunshot wound to temple without stellation. Courtesy Dr. Patrick Besant-Matthews. Used with permission. (See color insert following p. 202.)

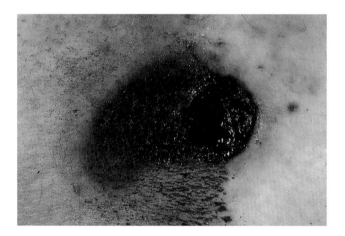

Figure 2.40 Near-contact gunshot wound with soot in wound track and some escaping to side. Courtesy Dr. Patrick Besant-Matthews. Used with permission. (See color insert following p. 202.)

skull; see Figure 2.41) create a much greater pressure and result in a much greater deformity than would occur in a soft, shock-absorbing cavity such as the abdomen (see Figure 2.42).

In areas such as the head, part of the gas is deflected by the bone. The result is a rapid expansion and outward tearing of the skin as the gas escapes confinement. The skin is elevated and expanded by the gas, slamming back against the muzzle of the gun. This can produce an imprint of the muzzle's end on the skin surrounding the perforation made by the projectile (see Figures 2.43, 2.44, and 2.45). This outward stretching and tearing of the skin

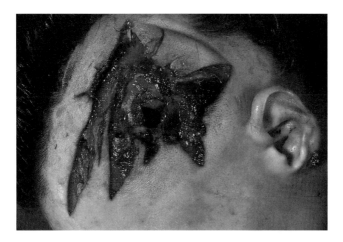

Figure 2.41 Large stellate-shaped gunshot wound deformity of head. Courtesy Dr. Patrick Besant-Matthews. Used with permission. (See color insert following p. 202.)

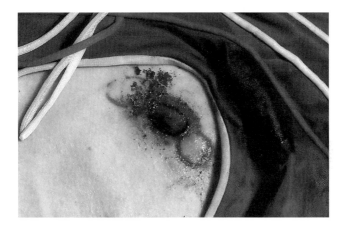

Figure 2.42 Contact gunshot wound to abdomen. Courtesy Dr. Patrick Besant-Matthews. Used with permission. (See color insert following p. 202.)

creates a wound with a *stellate* or star-shaped appearance (see Figures 2.46, 2.47, 2.48. and 2.49).

Because there are great variations in ammunition, and because target areas may or may not have underlying bone, contact range gunshot wounds vary greatly in appearance. These wounds range from a simple perforation to a large, complex skin defect; contrast Figure 2.41 with Figure 2.43.

Close range gunshot wounds occur when the range of fire is only a few inches. The dustlike residue of burned gunpowder known as *soot* is deposited on the target (see Figure 2.50). This soot creates an appearance known as

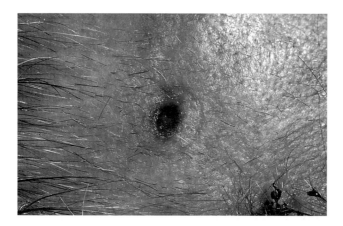

Figure 2.43 Contact gunshot wound to abdomen with light impression of muzzle. Courtesy Dr. Patrick Besant-Matthews. Used with permission.

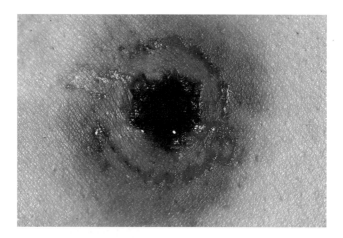

Figure 2.44 Very close to contact gunshot wound. Note impression of muzzle on skin. Courtesy Dr. Patrick Besant-Matthews. Used with permission.

fouling. The presence of soot usually indicates that the muzzle was less than 6 to 8 inches from the target* (see Figure 2.51).

Intermediate range gunshot wounds occur when the large particles of unburned or partially burned gunpowder strike the skin and create individual small abrasions or hemorrhages called *stippling.* The presence of stippling and the absence of fouling usually indicate a range of fire greater than 6 to 8 inches but less than 18 to 36 inches. As the muzzle-to-target distance increases, the zone of stippling becomes larger and its density diminishes.

---

* The term powder burn is a misnomer and should not be used. The heat and pressure of burning gases may create a skin reaction and gunpowder residues are deposited on or imbedded into the skin.

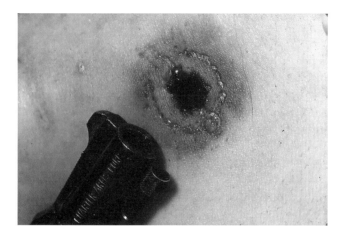

Figure 2.45 Contact gunshot wound to soft tissue showing barrel of gun. Courtesy Dr. Patrick Besant-Matthews. Used with permission.

Depending on the density of particles and the amount of hemorrhage produced, stippling can be obvious or subtle. *Always look for evidence of stippling and/or fouling around gunshot wounds* (see Figures 2.52, 2.53, and 2.54). Gunshot wounds should not be cleaned before photographing as evidence of soot deposits may be lost.

Distant range gunshot wounds occur when the muzzle of the weapon was greater than 18 to 36 inches from the point of entrance. Distant range gunshot wounds are usually round with an abraded margin. The defect in the skin is usually smaller than the bullet diameter due to the skin's ability to stretch. Because the bullet itself will have some soot on its surface that

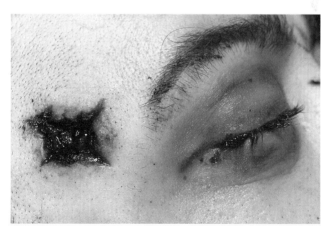

Figure 2.46 Stellate gunshot wound to temple and black eye. Note smaller skin deformation due to proximity to thinner bone of temple and blood traveling into eye socket causing "bruising." Courtesy Dr. Patrick Besant-Matthews. Used with permission.

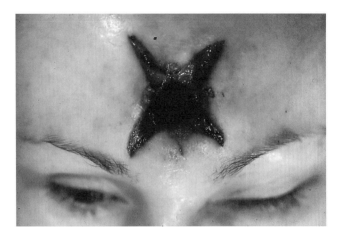

Figure 2.47 Stellate gunshot wound to forehead. Courtesy Dr. Patrick Besant-Matthews. Used with permission.

soot will be left on the skin as it enters. This is called *wipe off* and appears as a discoloration of the abraded ring at the point of entrance (see Figures 2.55 and 2.56).

A good illustration of range of fire would be when an individual is brought to the emergency room with a self-inflicted shotgun wound to the abdomen without any evidence of soot or stippling. It is impossible for an individual to hold a shotgun far enough away from the abdomen so that no soot or unburned powder would impact the skin. This scenario is immediately suspicious, and unless some device was used to hold the gun allowing it to be fired from a distance, the story does not match the injury presented.

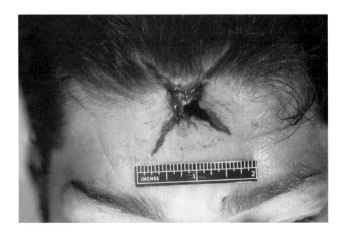

Figure 2.48 Perforating stellate gunshot wound to forehead with soot and ruler. Courtesy Dr. Patrick Besant-Matthews. Used with permission.

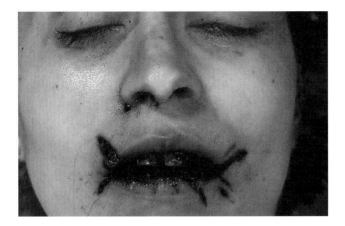

Figure 2.49 Stellate gunshot wound to mouth with soot deposit and stellate tearing. Courtesy Dr. Patrick Besant-Matthews. Used with permission.

Angle of Entry

This is the *relative angle at which the bullet entered the body*. The term *relative* here is important. Relative angle relates to the angle between the firearm and the target at the time the projectile penetrated the skin. The relative angle differs with the victim's position (standing, sitting, kneeling, etc.). A gunshot wound can provide information about the relative angle at which the bullet entered the body and associated conditions such as the type of firearm, ammunition, clothing, or interposing objects. When a bullet approaches the skin at an angle, the wider margin of abrasion on one side of the wound usually indicates the direction of travel (see Figures 2.57 and 2.58). Note the bullet itself just under the skin below the elbow. This angle helps detectives

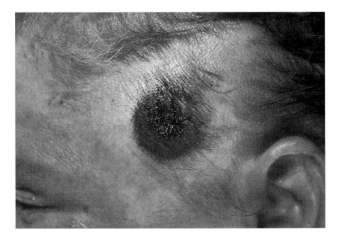

Figure 2.50 Contact gunshot wound to temple with soot deposit. Courtesy Dr. Patrick Besant-Matthews. Used with permission.

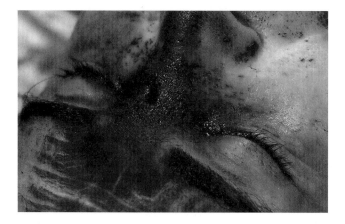

Figure 2.51 Contact gunshot wound to forehead with soot and stippling. Courtesy Dr. Patrick Besant-Matthews. Used with permission. (See color insert following p. 202.)

confirm the location of the shooter and may also help confirm or disprove different versions of the story (see Figure 2.59).

Types of Projectiles and Intervening Objects

Information about the type of weapon and ammunition used, or about any objects that may have been interposed between the firearm and the victim at the time of shooting, may be detected by careful observation of an entrance wound.

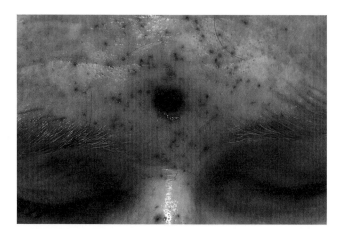

Figure 2.52 Small entrance gunshot wound to forehead with stippling, raccoon eyes, abraded ring. Courtesy Dr. Patrick Besant-Matthews. Used with permission. (See color insert following p. 202.)

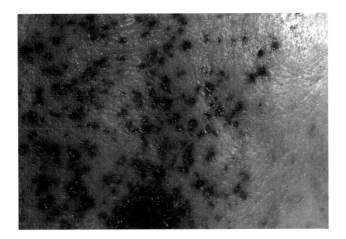

Figure 2.53 Stippling. Courtesy Dr. Patrick Besant-Matthews. Used with permission.

A wound from shotgun pellets has a different appearance than a wound made by a single shell bullet. The size, number, and velocity of projectiles will make a big difference in the severity of injury. Correct documentation of gunshot wound details (including corroborating photographs) is important.

Unusual or atypical entrance wounds can be produced for a number of reasons. A defective firearm or bullet may be the cause. Ricocheting causes a bullet to tumble, changing the appearance of the entrance wound. If a bullet passed through an intermediate object, it may carry traces of that object with it, embedding them in the subsequent object it hits. In the case of interposing objects, you may see bits of clothing, plaster or wood from a wall, or other

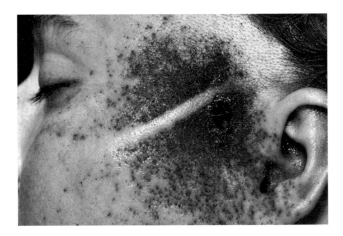

Figure 2.54 Gunshot wound anterior to left ear with fouling and stippling. Note sparing around sidepiece of glasses. Courtesy Dr. Patrick Besant-Matthews. Used with permission. (See color insert following p. 202.)

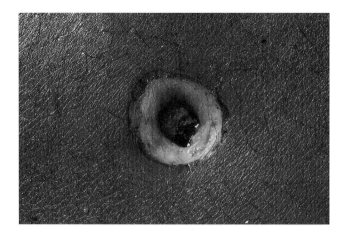

Figure 2.55 Abraded ring without significant soot deposit or wipe off. Courtesy Dr. Patrick Besant-Matthews. Used with permission. (See color insert following p. 202.)

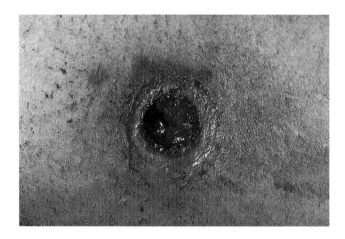

Figure 2.56 Abraded ring with wipe off. Courtesy Dr. Patrick Besant-Matthews. Used with permission.

signs that something else was hit before the projectile penetrated the victim's skin. Bullets that strike the ears, nose, or hands also produce unusual patterns due to contours in the skin.

Wounds on the skin can be difficult to interpret, especially when intervening items such as clothing, are struck first. For this reason, evidence, particularly trace evidence may be more easily found in clothing.* Depending on the tightness of the fabric weave and the numbers of layers that are perforated,

* Screen doors, regular doors, windows, and walls are examples of items that may be struck by a bullet before it hits a person. In the healthcare setting, clothing is the most common and can have significant evidentiary value.

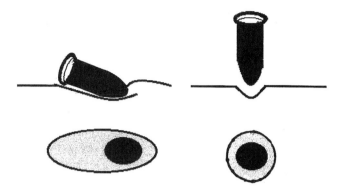

Figure 2.57 Bullet angle of entry sketch.

clothes may act as a filter and can retain most or all of the powder residue (see Figures 2.60 and 2.61).

Entrance and exit wounds *vary widely in size and appearance and can easily be confused.* Exit wounds are usually nonabraded lacerations. They may be larger than entrance wounds, but this is not always the case. Exit wounds may be circular, oval, and/or smaller than the entrance wound. This can cause confusion and misinterpretation. It is best not to label a gunshot wound as *entrance* or *exit*. Describe its size and characteristics instead. The use of terms such as *abraded ring* or *stellate-shaped* is better technique.

Skin offers greater resistance to perforation by bullets than any other tissue except bone and teeth. Most bullets that do not exit are lodged either in bone or are found just beneath the skin. Subcutaneous bullets usually present as small bumps under the skin. They are easily palpable and are often surrounded by a bruise.

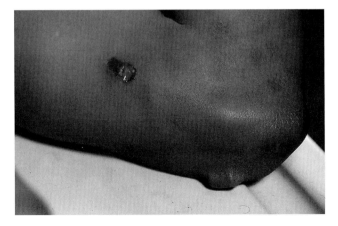

Figure 2.58 Angled entry with bullet under skin near elbow. Courtesy Dr. Patrick Besant-Matthews. Used with permission.

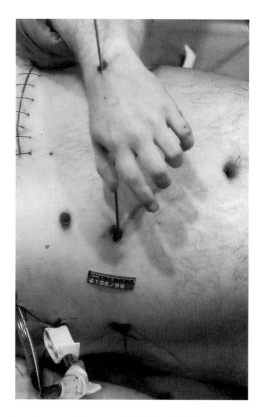

Figure 2.59 Angle of entry. Courtesy Dr. Patrick Besant-Matthews. Used with permission.

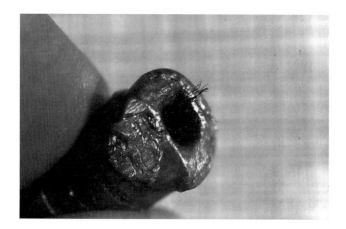

Figure 2.60 Spent bullet with fibers embedded in nose. Courtesy Dr. Patrick Besant-Matthews. Used with permission.

Figure 2.61 Near-miss gunshot wound. Note angle of entry and abrasion from bullet. Courtesy Dr. Patrick Besant-Matthews. Used with permission.

Photographing Gunshot Wounds

All wounds should be photographed both before and after treatment. Two photographs should be taken of each gunshot wound: one with a ruler and one without a ruler. Descriptive notes and an accurate diagram should always accompany photographs. (See the "Photography" section in this chapter.)

Important Terms to Remember about Gunshot Wounds

Soot is the product of burned powder and usually only travels 6 to 8 inches before dissipating. Soot that has been deposited on a target may be wiped off; do not clean a gunshot wound before photographing and documenting the presence of soot! If the muzzle is less than 3 inches from the target, the soot is compact and dense and it may not be possible to completely remove it.

Stippling is caused when grains of unburned powder are imbedded into the skin. This leaves a tattoo-like appearance near the entrance.

Searing is the result of hot gases escaping the barrel of the gun.

Stretching or *cavitation* is the formation of a cavity due to the transfer of kinetic energy from the bullet to the surrounding tissue. Contact between the weapon and the skin causes unique tissue damage due to the gases released *in addition* to the damage caused by the bullet itself.

Bite Marks

Bite marks are a significant injury pattern because of their evidentiary value. Their characteristics can be unique enough to match a bite mark to the dentition of the individual who made the bite. (See discussion of bite marks in Chapter 3.)

Human bite marks are oval (elliptical), superficial abrasions and may be single or double. They show indentation, puncture and/or petechiae (pinpoint bruising; see Figure 2.23), contusion (larger bruising), and abrasions. Bites from children are generally smaller than those of adults. Animal bites are oblong, V-shaped deep lacerations, and are almost always multiple.

Fractures of Bone

Bone may fracture in different ways depending on the amount of force and the manner in which that force was applied. Bone fractures display two basic patterns depending on the type of force applied.

- A transverse or V-shaped fracture indicates a direct force
- A spiral fracture demonstrates a twisting force

Again, documentation of fractures should be limited to their existence, appearance, and placement, leaving interpretation to the orthopedic or forensic specialist.

Thermal, Electrical, and Chemical Injuries

Healthcare providers encounter burns in association with fires, heated objects, chemical agents, electrical accidents, and lightning strikes. Electrical burns are typically incurred by construction workers and electricians. There are, however, cases among do-it-yourselfers who fail to observe prudent practices associated with electrical wiring, appliances, and tools. Electrical burns are usually the result of accidents or intentional acts of homicide. However, there are rare reports of suicide involving electricity. Burns from these causes differ significantly from those induced by lightning. They are the combined result of direct tissue heating, contact burns, arc burns, and thermal burns from the ignition of clothing (Stewart, 1990).

Forensic investigators are responsible for gathering evidence and reconstructing the scenario to help determine whether or not there is civil or criminal liability, or whether it is merely the result of an accident. Healthcare personnel can make observations, document, identify, and collect evidence that provides invaluable information to law enforcement, death investigators, arson investigators, and other healthcare providers.

Thermal Burns

Classifications

Burn wounds are classified in terms of the extent of tissue damage. Historically, burns were placed into four groups based on the degree of injury.

- First-degree burns: Involve the epidermis; victim experiences redness and pain
- Second-degree burns: Affect both the epidermis and dermis; blister formation is common
- Third-degree burns: Extend into subcutaneous tissue
- Fourth-degree burns: There is deep tissue loss including muscle and bone

As a practical matter, most wounds have areas with more than one classification. A wound may have deep tissue involvement in the center, but first- and second-degree injury at the periphery. Recent burn literature suggests that burns fall into two categories:

- Partial thickness: superficial or first- and second-degree burns
- Full thickness: characterized by whitish skin changes, charring, or deep tissue loss (fourth-degree burns)

It is essential to examine the burn wound for patterns that suggest direct contact with a heated object such as a lighted cigarette or curling iron. These objects can readily produce full-thickness burns, especially when intentionally held in contact with the skin. Defenseless young children and the elderly are the most at risk for these types of burns because their skin is more sensitive to temperature extremes and they are less able to protect or defend themselves. Accidental burns have a more erratic appearance because the victim is able to rapidly pull away from the hot object (Olshaker, Jackson, and Smock, 2007).

Scalding Burns
Scalding burns associated with hot liquids or steam can be divided into three groups:

- Immersion burns
- Splash or spill burns
- Steam burns

Although any of these burns may be accidental, it is essential to evaluate each situation to detect intentional acts of abuse or torture. When bodies or extremities are immersed forcefully, the burns are often well demarcated. When the burn is accidental, the uneven margins suggest that splashing has occurred. The irregular appearance suggests that the victim had the freedom to move willfully away from the hot liquids. In extremely hot water (130–140°F), contacts as brief as one second can cause full-thickness burns.

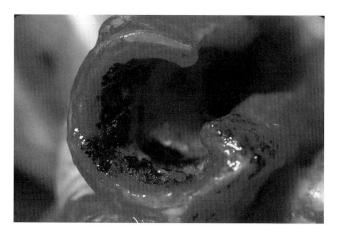

Figure 2.62 Cross-section of trachea with charred areas. Note the classical pink color of high carbon monoxide exposure. Courtesy Dr. Patrick Besant-Matthews. Used with permission. (See color insert following p. 202.)

Inhalation Injuries

Patients with thermal burns are at high risk for additional inhalation injury (see Figure 2.62). Signs and symptoms of inhalation injury include:

- Shortness of breath and severe coughing
- Difficulty swallowing
- Hoarseness
- Soot on face or mouth
- Singed hair on the head or face, especially around the nose or mouth
- Facial burns

Respiratory involvement associated with a burn suggests the victim was unable to leave a confined space or that an explosive substance or accelerant may have been associated with the fire. All observations and other information should be documented carefully and conveyed to an arson investigator or other forensic specialists.

Important evidence to preserve includes:

- Photographs of the scene, the victim, and any obvious burn injuries
- All clothing associated with the victim or others, including discarded jackets, gloves, shoes, or other pertinent items found at or near the scene
- Any secretions from the mouth or respiratory tree including blood, sputum, fluids from bronchial washings, and nasal or oral swabs

Electrical Burns

Definitions

- *Electrocution* is death by electrical shock.
- *Electrical shock* is the passage of electrical current through the body with or without subsequent death.

When an injured or dead victim is found near electrical wires, electrical shock or electrocution should be suspected.

Basics of Electricity

An *ampere* is the amount of electrical current (volume or flow of electricity) passing a given point in an electrical circuit. Contrary to common assumption, circuit breakers are not designed to prevent shock to humans or other living organisms; they are designed to stop a current overload, thus preventing a fire. Circuit breakers are usually rated at 15 amperes or more. The amount of electricity needed to affect a human is in milliamperes (a milliampere is one one-thousandth of an ampere). One milliampere can create a slight tingling sensation in humans. In 90% of people, the amperage required to create tetanic contracture (the *no-let-go* phenomenon) is between 0.009 (9 milliamperes) and 0.015 amperes (15 milliamperes) (see Figure 2.63).

In victims with previous burns, or nonintact skin, less amperage is required to create organ and tissue damage. Even at very low amperages of electrical shock current, there is a lower probability of complete recovery due to multiorgan damage because the electrical current often follows the vascular, lymph, and nervous system and can cause permanent damage to them.

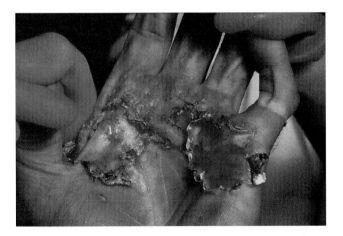

Figure 2.63 "Blow out" appearance of electrical burn to hand. Courtesy Dr. Patrick Besant-Matthews. Used with permission.

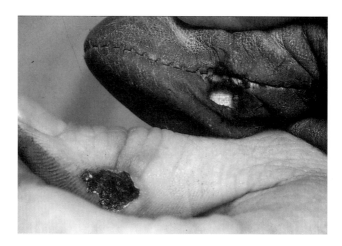

Figure 2.64 Electrical burn through glove to thumb. Courtesy Dr. Patrick Besant-Matthews. Used with permission.

Even if a suspected victim of electrical shock may seem fine at the moment, it is imperative that they seek medical treatment immediately. Why? Because the tissues have been exposed to a damaging force sufficient to silently injure or destroy cells. Swelling, airway obstruction, and/or cardiac complications can follow. Monitor your patient continuously for signs of cardiac involvement, especially V-fib or asystole. These patients can deteriorate precipitously. Awareness of the patient's condition is extremely important because the patient's condition may require prescience (anticipation of potential patient deterioration) and rapid intervention.

Physiologically, flexor muscles are stronger than extensor muscles. This is why the no-let-go phenomena occurs. The electrical stimulus causes the flexor muscles to override the extensors, making it physically impossible for someone to let go of the active electrical circuit.

The effects on tissue are dependent on two things: time and current (amperage). The determining factor in shock cases is the amount of current to which the victim is exposed. The amount of electrical resistance (clothing, shoes, environmental conditions) presented by the shock victim will influence the amount of current reaching the body; that is, the greater the resistance (protection), the less current that is allowed to flow into the body. Some clothing can act as a poor insulator against electrical current and can allow electrical current to pass through relatively easily, depending on the resistance of the material encountered (see Figure 2.64). It does, however, provide some protection. Metal is a good conductor of electricity and heat. In clothing, it can provide the point of entry into the body (see Figure 2.65). Note the overall effect (see Figure 2.66). Note the general resistance provided by the thick leather sole, but the easy entry point where metal was attached (see

Figure 2.65 Belt buckle point of electrical entry, left upper corner. Courtesy Dr. Patrick Besant-Matthews. Used with permission.

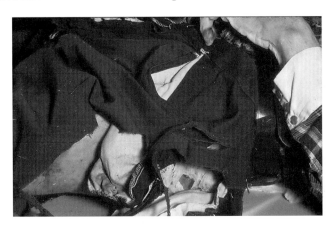

Figure 2.66 Effect of electrical contact through belt buckle. (Note belt below and to left of cuff of sleeve.) Courtesy Dr. Patrick Besant-Matthews. Used with permission.

Figure 2.67). It is for this reason that steel-toed shoes are not recommended for those working with electricity or on fires.

It is important both for the first responder and follow-up EMS and hospital personnel to understand that the amount of current required for serious electrical shock injury is about one thousand times *less* than it takes to trip a breaker in a building. It only takes 20 to 40 milliamps (0.020–0.040 milliamps) to paralyze the intercostal muscles and diaphragm, leading to respiratory arrest. This level of current may not yet be high enough to cause cardiac symptoms, so the patient's heart is still beating normally. Beginning cardiopulmonary resuscitation (CPR) as soon as possible is vital because one or two rescue breaths may be all that is needed to stimulate spontaneous respiration.

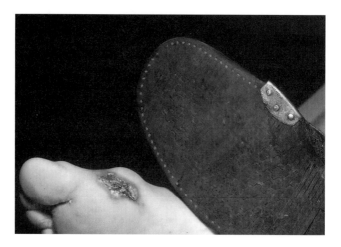

Figure 2.67 Point of electrical entry through metal attachment to shoe. Courtesy Dr. Patrick Besant-Matthews. Used with permission. (See color insert following p. 202.)

If spontaneous respiration does not stimulate the patient to breathe on his or her own, continue ventilating the patient and watch for signs of cardiac involvement.

Key Point:

In the case of electrical shock injury, it is not safe for responders to begin CPR until the patient is away from potentially energized electrical wires or circuits.

Tissue damage is largely due to heat, which induces vascular spasms, thrombosis, neurological injury, and muscle necrosis. Certain body organs and tissues are more subject to electrical energy than others due to their physiological properties, which contribute to low resistance. For example, central nervous system tissue, which is specifically designed to carry electrical signals, has exceptionally low resistance. Cardiovascular and respiratory systems also have low resistances due to their high composition of electrolytes. Skin, tendon, fat, and bone have higher resistances, but if the voltage is high, the heating becomes greatest in the more electrically resistive tissues. During prolonged contact with electrical energy, fat and tendons actually melt and bone incurs significant periosteal damage. (Stewart, 1990). Other damage includes muscle necrosis and rhabdomyolysis. Rhabdomyolysis can be detected by massive amounts of hemoglobin in the urine, but without red cells or red cell fragments. Assays for myoglobin should also be done along with creatinine phosphokinase (CPK) levels. Levels range up to 20,000 units for victims of severe electrical injuries.

Key Point:
Remember the ABCs! One or two rescue breaths may be all that is needed
for the shocked patient who is not breathing but who has a maintained a
pulse.

Ventricular fibrillation is a common symptom of electrical shock.
Asystole (no heartbeat) can occur following prolonged shock or shock with
higher current levels. Between 50 and 150 milliamperes may cause ventricular fibrillation. Seventy-five milliamperes may be sufficient to cause cardiac
arrest. Even at these very low amperages, there is a low probability of complete recovery due to multiorgan damage because the electrical current often
follows the vascular, lymph, and nervous system.

Electrolyte imbalance, release of myoglobin into the bloodstream, scar-
ring, and/or eventual cellular death can be long-term consequences of having been shocked. For instance, rhabdomyolysis releases myoglobin into the
bloodstream. The myoglobin molecule is larger than the kidney can normally
handle, plugging the arterioles and causing damage to the kidneys, even to
the point of kidney failure. The history of this event would be the clue that
high volumes of IV fluids are needed to help flush the kidneys.

Electrical burns can create extensive local burn wounds as well as a
range of systemic effects, depending upon the intensity of current flow
and duration of contact. Skin and tissue resistance and current pathway
through the body are also determining factors in the electrical injury and
the resultant burn wounds. Factors that increase current density include
perspiration, tissue resistance, the electrical current pathway through the
body, electrical grounding, and a small contact area and high voltage (more
than 1,000 V).

Types of Electrical Burns
There are three types of electrical burns: *contact burns, arc-flash burns*, and
arcing burns.

Contact Burns   Contact burns occur when the body comes into direct
contact with energized electrical appliances or current. Electrical contact
wounds (formerly referred to as *entry* and/or *exit wounds*) may or may not be
present depending on length of contact time, condition of the skin surface
affected (wet vs. dry, thin vs. calloused), and the amount of current passing
through the body. Assess these issues during the primary assessment (see
Figures 2.68 and 2.69).

Healthcare providers should be alert to lip or mouth burns in toddlers
who bite or suck on electrical cords or sockets. The electrolyte content of
saliva rapidly conducts the electricity and an oral burn occurs. Serious
sequelae from these burns are uncommon, but they deserve a thorough

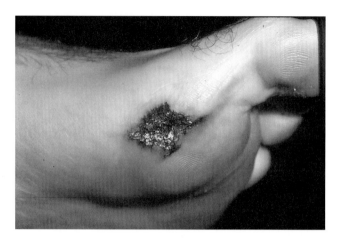

Figure 2.68 Electrical burn on foot. Courtesy Dr. Patrick Besant-Matthews. Used with permission.

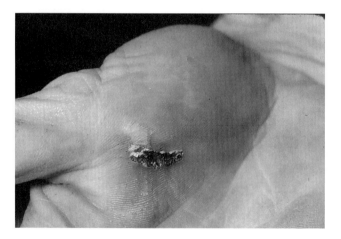

Figure 2.69 Electrical burn at juncture of thumb and palm. Courtesy Dr. Patrick Besant-Matthews. Used with permission.

initial assessment and follow-up. Management by a surgical specialist may be needed to prevent lip contractures and to ensure optimum cosmetic results as healing occurs.

Electrical Damage to Vital Organs* AC electrical energy below 1,000 V produces ventricular fibrillation at a rate of 3,600 times per minute. The human heart can only tolerate rates up to 300 beats per minute, above which

---

* A standard laboratory blood test (CKMB) should be performed on all suspected victims of electrical exposure. It measures the level of red blood cell destruction. Normal ranges differ, but usually range from undetectable to 7U/L or so (http://www.clinlabnavigator.com).

the patient promptly expires. High-voltage electrical energy, however, essentially overrides the heart's natural electrical conduction system. Once the electrical flow has been stopped, the heart can resume a normal rhythm (James and Nordby, 2009). The environment and historical accounts of witnesses are important in determining whether a death has been caused by electrocution. Deaths from high-voltage electricity produce obvious burn injuries. Deaths from low-voltage electricity may produce no evident physical injury at all. The death scene should be carefully photographed and all clothing should be saved using standard evidence collection techniques. Burns to the skin can result from clothing that has caught on fire.

Where were the contact points? Assessing potential vascular, lymph, muscle, and nerve pathways that may have been involved will assist in evaluating possible internal and organ damage. Cardiopulmonary involvement should be suspected when the contact is between one hand and the other. Hand- or foot-to-head is highly suspicious for brain involvement, and may result in brainstem damage. The patient's age, basic general health,* and mental state are big factors in how well a patient recovers from an electrical shock.

Key Point:
After an electrical shock incident, in addition to providing immediate care for the victim, a thorough account of the accident should be obtained in order to estimate the extent of nonvisible tissue injury.

Assess the patient for superficial burns and treat according to local protocol. Assess the superficial burn area using the *rule of nines* or other standard method of estimating the amount of body surface area involved.

The *rule of nines* is a method of assessing the percentage of body surface burned using multiples of 9%. (Emedicinehealth, 2010).

Head = 9%	Each arm = 9% (4.5% each side)
Chest – front = 9%	Each leg = 9%
Abdomen = 9%	Palms and groin = 1% each
Upper and mid-back = 9%	Lower back and buttocks = 9%

---

* Younger patients with healthy hearts have a greater potential for recovery with EMS intervention. Older patients with coronary artery disease, lymph system diseases, nervous system diseases, and possibly autoimmune diseases involving any of the above systems or musculoskeletal diseases may not fare as well as younger patients.

Arc-Flash Burns   Electrical arcing is the "luminous discharge of electrical current through the atmosphere" (Witter, January 29, 2010).*

An *arc-flash* burn is an electrical burn caused by radiant heat, not because of direct contact with an electrical current (as when trying to pull a meter). The severity of burn is dependent upon the temperature of the arc, duration of exposure, and distance from the arc. Burns can range from mild skin reddening to full-thickness third-degree burns. Temperatures involved can be several thousand degrees and may last as little as 0.1 second. An exposure of 0.1 second at a distance of 2 feet can cause third-degree burns. Up to 80 % of electrical burn injuries are the result of arc flash and ignition of clothing.

Arcing Burns   Arcing burns are a combination of current flow through the body and a flash of electrical arc. The victim may experience all injuries associated with a contact burn plus those of an arcing burn. Loss of limb(s) is very possible.

Treat the patient under standard burn care protocol and continue to assess for inhalation burns. Be prepared for tracheal edema and respiratory arrest.

Arc Blast Injuries   An *arc blast* is a pressure wave caused by the rapid expansion of gases and conducting material with flying molten materials. It is a violent explosion of electrical components and rapidly moving shrapnel. A 25,000 ampere arc blast can exert 480 pounds of force on the average person's body at a distance of 2 feet. The blast can destroy structures, and knock workers from ladders or across a room.

First responders and subsequent healthcare personnel should maintain a high index of suspicion for concussion, skeletal and/or spinal injuries, flash burns, and penetrating injuries. Ultimate survival from these injuries is low. *Scene safety is paramount in arc-blast situations.*

Lightning Injuries and Death
Lightning is a high-current electrical injury that can kill by direct or indirect contact.

Of greatest concern with an indirect lightning strike is the presence of differing voltages in the ground. Think of a stone thrown in a lake. The ripples move away from where the stone hits the water and the ripples eventually die out. The same thing happens with lightning. When lightning hits ground or something on the ground, the voltage gradients move away from the point of the strike, and the voltages in the ground

---

* Interestingly, arcing can result from an indirect lightning strike with the current conducted over a metallic path and is a source of fires in many homes that use a newer form of gas piping known as corrugated stainless steel tubing. If the metallic components, water pipes, and gas pipes are not all electrically bonded together, there can be an arc caused by an indirect lightning strike that can burn a hole in the gas line and ignite a fire.

vary from one point on the ground to another. This phenomenon occurs instantaneously. If someone is near an indirect strike, their feet can be subjected to differing voltages in an extremely short period of time, which can be lethal.

Direct lightning strikes often result in immediate unresponsiveness and death from cardiac arrest or motor paralysis involving the respiratory centers. In these cases, thousands of amperes flow through the body. The cardiovascular system is immediately compromised by massive vasomotor spasm causing loss of peripheral blood flow, loss of sensation, and loss of color in the extremities. Peripheral arterial thrombosis and tissue death may follow. However, other victims are merely stunned or have retrograde amnesia. Paralysis or slow reflexes may be noted, and it is not uncommon for victims to be unable to speak or to hear. Hysteria, personality changes, and visual impairment have also been reported. Typically, if the individual survives, he or she may suffer less brain damage than other individuals experiencing respiratory or cardiac arrest because cell metabolism is halted as electrical forces pass through the body. Myoglobinuria is common with lightning-related injuries. A simple urine test helps to confirm that an injured individual has been struck by lightning or suffered a similar electrical energy event* (Lanros and Barber, 1997).

Lightning should be considered as a cause of death when an individual is found outside and there is, or has been, a thunderstorm in the area. Featherlike imprints are characteristic of being hit by lightning strikes (see Figure 2.70). As electrical current spreads over the skin, it produces linear, spidery arborescent and erythematous skin imprints and discolorations. This is a hallmark of electrocution. These wounds should be promptly photographed because

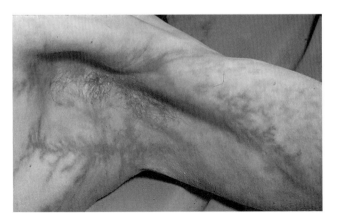

Figure 2.70 Featherlike appearance of lightning injury. Courtesy Dr. Patrick Besant-Matthews. Used with permission. (See color insert following p. 202.)

* Myoglobin is not normally found in urine.

this characteristic pattern may disappear within hours of the death (Wagner, 2009). If lightning exposure is suspected, there should be a thorough search for exit wounds, paying particular attention to obscure areas such as the bottom of the feet or the anus. The presence of lens injuries and perforated tympanic membranes also suggest lightning injury. Lightning strikes often result in rapidly developing cerebral edema, which can be found on autopsy.

Weather-Related Hazards

Overheating-Related Tissue Injury

Emergency responders and forensic personnel are often subject to extremes of heat and cold at the scene of an incident. Working fast and furiously with heavy protective clothing can contribute to both increased heat production and the inability to dissipate the heat generated. Individuals tend to sweat, losing valuable fluid from the body. For well-acclimatized individuals accustomed to working in hot environments, tolerance will be significantly better than for those workers who typically perform in comfortable, air-conditioned spaces. Those in the latter group fall prey more readily to atmospheric temperature extremes. Other predisposing factors are certain medications, high humidity, obesity, and cardiovascular disease.

Heat-related problems, in order of their severity, include:

- Heat cramps: painful contractions of skeletal muscles due to sodium loss from sweating. Fluid replacement is the mainstay of both preventing and managing this problem. Sports drinks are recommended since they contain important electrolyte replacements.
- Heat syncope: fainting episode due to vasodilation, peripheral pooling of blood, volume deficits, and sluggish vasomotor tone. Fluid and electrolyte replacement and rest in a recumbent position are required for recovery.
- Heat exhaustion: a more serious heat-stress condition that encompasses heat cramps, heat syncope, and an altered mental status. Headache, dizziness, irritability, a rapid heart rate, and hyperventilation also commonly occur. The treatment is the same as for heat syncope.
- Heat stroke: the most serious heat-related condition resulting in death if not vigorously treated. The hypothalamus loses its ability to regulate body temperature. Normal modes of temperature regulation such as sweating are simply overwhelmed. Seizures may occur and the body's basic metabolic and cardiopulmonary systems fail. Life support measures and prompt definitive emergency medical care at a hospital are required.

Cold-Related Tissue Injury

Any prolonged exposure to cold resulting in significant vasoconstriction and sluggish circulation can create injury to exposed areas of flesh. Eventually, a person's core temperature is affected, leading to death.

Cold-related problems, in order of their severity, include the following.

Frostnip   Frostnip is a condition in which exposed skin areas farthest from the trunk are nipped (nose, ears, cheeks, chin, hands, or feet) from cold exposure. The victim experiences burning and tingling sensations.

Frostbite   Frostbite is a condition that occurs when frostnip persists and progresses to the extent that the superficial layers of the skin are frozen. Deeper tissues usually remain resilient, however. The skin appears waxy and is numb.

Upon rewarming, the affected areas become mottled or purplish and remain numb. Edema, burning, stinging, and blister formation eventually occur. Deep frostbite damages not only the skin, but the subcutaneous tissue and blood vessels as well, producing pain and throbbing that persists for several weeks. When the blisters eventually dry and the skin sloughs, the new skin will remain sensitive to cold. Pain, itching, and excessive moisture are felt throughout the affected area. In the most severe cases when management is delayed or poorly executed, gangrene may complicate the condition, requiring amputation.

Hypothermia   Hypothermia is a life-threatening condition that occurs when the core body temperature is less than 95°F or 35°C. The most serious problems occur when the body temperature reaches 90°F or 32.2°C. At this temperature, shivering stops and muscles become rigid. When the core temperature falls below 78°F or 25.5°C, death is typically imminent. An important fact to remember is that the brain and vital processes that ordinarily fail when anoxia and circulatory arrest occur are somewhat protected because of the slowing metabolic demands that occur during hypothermia. If an individual is found in a cold environment and appears rigid, resuscitation should be attempted. Successful recoveries have been reported up to two or three hours after what was apparent death. Therefore, the American Medical Association states that "the recommended treatment of hypothermia in the field is core re-warming to prevent post-rescue collapse" (JAMA, 1992; Alaska protocol, 2010; U.S. Coast Guard protocol, 2010).* Some authorities believe that you cannot

---

* In dire circumstances where warming equipment and supplies are not rapidly available, wrapping the victim together with two undressed normothermic individuals has proven effective.

declare death unless rewarming has taken place and resuscitation has been attempted.

Chemical Burns

An important aspect of chemical exposures is that caustic materials continue to burn until they are removed or neutralized. The pathophysiological changes are different than those from thermal burns. The chemical reaction may be an oxidization or reduction reaction, or a protoplasmic poison, desiccant, vesicant, or corrosive (Stewart, 1990). The skin destruction is a product of the extent and nature of the exposed skin surface and the length of time exposed. Delicate areas such as the face, eyes, and genitalia are often the target in cultural crimes aimed at disfiguring the victim or rendering the victim to a state of permanent discomfort or dependency.

The eschar provides some clues to the chemical used. Sulfuric acid causes green-black to dark brown discoloration; nitric acid produces yellow eschar and tissue staining; hydrochloric acid creates yellow-brown eschar and tissue staining, and hydrofluoric acid produces a grayish to brown eschar (Stewart, 1990).

It is rare that chemical burns result in death. However, there are incidents where individuals are targeted and intentionally assaulted with chemicals in an attempt to maim or kill. Healthcare providers should be attuned to historical details related to the incident and the scene in order to determine if foul play is a factor to be considered. It is rare that chemicals alone are used to commit homicide. Some cultures use chemicals as a form of punishment. Most of these cases involve the face, hands, or genitalia.

In Conclusion

The healthcare provider must have an appreciation of the basic issues in tissue injury due to heat and cold, burns or electrical injuries, including lightning strikes. Cold-related injuries, especially of the limbs and extremities, often mimic burns. These thermal injuries, as well as hypothermia, pose special challenges for healthcare providers and forensic personnel.

First responders, clinical personnel, and death investigators must carefully assess patients to distinguish accidental injuries from intentional abuse or violent acts by offenders. Initial assessment of accidental versus nonaccidental causes and other forensic issues is a vital process, ensuring that victims of abuse are properly identified and protected, those responsible apprehended, and the innocent are not incorrectly branded as perpetrators. Precise collection of evidence and prompt reporting of forensic cases will greatly facilitate the decision-making process and apprehension of offenders.

Photography

There are many reasons to photograph a scene, a person, and/or evidence. Photographs document the location of a crime and confirm the presence or absence of evidence found at the scene. Photographs are critical for those not actually at the scene. They show the visual relationships among the scene, items at the scene, and the victim.

Photographs provide a visual record for witnesses and the police. At trial, photographs become a visual aid to the court and jury, adding weight and credibility to testimony.

It is important to remember that in sexual assault cases the *person* is the crime scene. If the patient is reluctant to be photographed, remind him or her that photographs cannot be taken later. Photographs can be taken at the present time and saved in case the victim wishes to prosecute at some time in the future. For this reason, it is frequently in the victim's best interest to be photographed.

Institutions and businesses should have a formal policy including forms for patient consent or refusal to be photographed.

Photographic Evidence

Photographs can aid in documenting evidence that is found at a scene or on the body of a victim or offender. It is especially valuable in recording scene details that might be easily overlooked while attending the injured, or to preserve the initial appearance of wounds that may heal or change over time such as bruises, abrasions, or lacerations. Photographs will ensure that other witnesses, caregivers, and members of the jury can appreciate the forensic scenario and its related evidence. Photographs visually preserve the scene and associated evidence, and illustrate relationships among the overall scene, various items of evidence, and the victim(s) or offender(s). Photographic images add credibility to testimony and permanently preserve recollections of events and circumstances that tend to be lost over time.

Photographs should be obtained promptly before any aspect of the scene is modified by law enforcement or medical intervention. Having stated that objective, healthcare personnel must never delay or defer lifesaving care merely to obtain evidentiary photographs.

Purposes of Photographs

The purposes of photographs are as follows:

- Documents evidence at the scene or on the body
- Records scene details that might be overlooked while providing care to the injured

- Preserves the initial appearance of wounds before treatment or healing processes begin
- Provides witness of jury members' access to scene details and on-site evidence
- Demonstrates relationships among the scene, evidence, and victims
- Supplements written medical records
- Can be used as a teaching aid for law enforcement, emergency care personnel, and forensic investigators

When a human being has been wounded, assaulted, or killed, the body is considered to be the crime scene. It is vital to explain the benefits of recording injuries as soon as possible, even though the victim may not have yet decided whether to press charges against the offender. It is preferred to have the photographs taken and to not need them, than to wish later that you had recorded these valuable evidentiary images. Bruises fade, scratches and abrasions fade, and other wounds heal. It is typically in the victim's best interests to consent to forensic photography. Each hospital or healthcare agency will have specific policies and procedures regarding forensic photographs, including forms for informed and implied consent.

Basics of Photography

Keep these points in mind:

- Photograph as soon as possible.
- Photograph as you find it. Do not disturb the scene or the victim before taking photos.
- Have a plan and stick to it.
- Get a complete series of photographs.
- Treat each exposure as if it contains a hidden clue.
- Record each photo in a log as it is taken. The log becomes a reference. This is especially useful in conjunction with use of a contact sheet.
- If you use 35-mm color slide film, not black and white, the slides themselves become evidence.
- Use point and shoot until you're comfortable with more sophisticated cameras.
- Digital photography is generally accepted both by law enforcement and by the judiciary. Understand any restrictions on taking, securing, and presenting digital photographs in your area.
- Look at lighting; consider the direction and intensity. When lighting is questionable, use a flash.

A photographic log should be maintained for each case with the following information:

- Identification data for the subject
- Date of the photography
- Name of law enforcement agency and the case number
- Medical record number
- Type of examination completed
- Identification of the photographer
- Chain of custody form
- Location of non-rewritable CD copy of images

Consent to Photograph

Consent is not required for forensic photographs if the subject or focus is

- Unconscious or deceased
- A public place such as a street, sidewalk, restaurant, or public event
- A forensic scenario or subject when there is a court order to obtain such photographs of the scenario and its victims or offenders

Consent is required for forensic photographs if the subject or focus is

- Able to provide informed consent (e.g., adult who is alert, oriented)
- Preserved for teaching purposes or publication

Written, informed consents are typically requested, although implied consents are also acceptable.

Informed Consent

The photographer must explain to the subject:

- The reason for requesting the photograph
- Risks and benefits of photographs
- How the photographs might be used

Note: Photographers must appreciate that subjects may refuse to provide consent. However, if the reasons for obtaining the photographs are fully explained, most subjects are willing to have images recorded. Many hospitals now obtain the consent for photography as a clause within the routine consent forms for care and treatment. A sample consent form is provided in Appendix 2.14.

Implied Consent

This type of consent is reserved for those individuals who are too injured, ill, or who are unconscious, and therefore unable to execute an informed consent, but who would be expected to consent if they were able to understand the circumstances. For example, it would ordinarily be in the subject's best interests to have photographic documentation that would help in the evaluation and treatment of their injuries or to aid in the prosecution of a perpetrator, offender, or other individual who might have liability for their condition.

Admissibility of Photographs

The photographs and the photographer must be able to withstand judicial tests for authenticity, integrity, and credibility. Courts must determine if the photograph:

- Shows original appearance and findings
- Fairly and accurately depicts what was visualized by eyewitnesses
- Assists in identification or characterization of wounds or other injuries
- Aids the court or jury in evaluating elements of the crime scene
- Is not unduly gruesome or intended to sensationalize facts or events

Courts usually challenge the photographer by requesting information such as:

- Exactly when and where the photograph was taken
- Why the photograph was taken
- What the photograph precisely depicts
- Type of camera used and whether or not a filter was employed on the lens
- If implied or informed consent was obtained from the subject
- The processing details of the image
- Whether the chain of custody can be verified

Qualities of Photographs for Evidentiary Purposes

The photographer must:

- Use a new memory card or a new roll of film for each particular case or subject.
- Record an initial image that provides *case-specific identification data* for the photographs to follow.
- Exclude bystanders when possible unless their presence might aid to authenticating the crime scene.

- Compose images to frame the original appearance of crime scene elements, including wounds or injuries, to highlight important evidence, and to demonstrate its relationship to other elements of the scene.
- Use a small ruler in the plane of body parts or wounds to demonstrate the exact wound size in inches or centimeters. An ABFO ruler is typically used for this purpose (see Figure 2.2, ABFO rule).
- Remove extraneous items from the photo such as bloody clothing, soiled dressings, or unused medical equipment.
- Ensure that recorded descriptions in the medical record correspond with what is illustrated in the picture. If there are discrepancies, these should be annotated and explained. (For example, if a hemostat is in place to arrest arterial bleeding in the initial photo, but was removed after the victim died and therefore is not present on a subsequent photo, this should be explained.)

Equipment for Forensic Photographs

There are several types of cameras that may be used, but in today's forensic community, digital images are preferred to film. First responders and healthcare personnel should concentrate on good-quality images that convey the evidence, and most digital cameras support that objective by offering autofocus, lighting correction, automated auxiliary flash, and other fail-safe devices. Even an amateur is capable of recording a useful image. All first responder units should have a digital camera with a new memory card in place upon arriving at any forensic scene. The paraphernalia associated with sophisticated 35-mm or other older camera types, which require film, flash attachments, filters, and other adjuncts, are too cumbersome to manage when your primary role is the provision of emergency care to injured victims or a suspect. The wide range of available options and prices make the digital camera an excellent choice for first responders and in-hospital applications. Individuals recording digital images must be aware of nuances about the operation of their camera including formatting the memory card, changing batteries, file compression precautions, and preserving photographs to ensure that they will be admissible in court.

Scene Photography

Ordinarily, the crime scene photos are taken by law enforcement, but may be taken by first responders from a fire department, emergency medical services, or others at the forensic incident. There are many specific elements to crime scene photography that are beyond the scope of this book. However, there are some fundamental rules that any forensic photographer should follow.

In the early years, there was considerable concern about the ease of altering digital images. It was thought that their authenticity might be questioned or that the potential for tampering would cause the images to be nonadmissible in a forensic case. However, there are several safeguards used today to prevent image manipulation and to detect any alteration that might have occurred.

Infrared, ultraviolent, and alternate-light-source photography is occasionally used to capture images of physical evidence that might otherwise be impossible to visualize or to successfully record with ordinary light sources or a flash.

Sequence of Photographs

Photographs are taken sequentially in a proscribed order starting at the perimeter of a scene or person and working toward detail.

The first photograph taken is known as the *distance* photograph. It is taken as far away from the scene/victim as is necessary to establish the location and general conditions. Subsequent photographs are taken increasingly closer to the scene/subject with the following guidelines in mind.

- An *intermediate distance photograph* records the overall scene and ties distance and close-up images together. At a scene, it places the victim within the scene and locates specific objects within that scene.
- The *close-up photo* establishes the victim's identity (face and other identifying information), records detail, the victim's position, and documents wounds. It is imperative that photographed subjects understand the importance of having photodocumentation to verify their identity.
- The *ultra close-up* of any detail (1:1 of specific wounds and items) should include a white balancing or gray scale to verify subtle aspects of coloration and lighting.

During the initial scene encounter, certain items of evidence may be collected. Each specific item should be photographed in place before collection and once again after collection to ensure that this process is fully documented.

Photographing Human Abuse and Assault Injuries

Victims of violence must provide written or informed consent prior to physical assessment and before obtaining photographic images of the condition of their bodies and the related injuries. Parents or legal guardians may consent for a minor child.

Physical abuse and injuries of interpersonal violence as well as other traumatic injuries create contusions, abrasions, lacerations, cuts, burns, punctures, and the characteristic wounds associated with various firearms. The characteristics and locations of such wounds or soft tissue defects provide helpful clues in determining injury forces or weapons used in an assault. When feasible, such photographs should be obtained before permitting the patient to wash hands or to shower and change clothes. The body and clothing often provide additional trace evidence and physical clues about the biomechanics of injury.

It is helpful to annotate body diagrams to accompany the series of documentation photographs (see Figure 2.1).

Photographs and body diagrams are to be managed as an element of the medical record and are subject to the same guidelines for safeguarding and release. Chain-of-custody forms are vital tools for verifying the authenticity and preventing tampering with images. See Chapter 3.

A photographic log should be maintained with the name of patients (or case number), an identification photo, the date and time that images were recorded, and the name of the photographer along with credentials and affiliation.

The human subject should be photographed with an orientation photo that reveals identification information and the anatomical locations of primary injuries. Intermediate range photos and close-ups should be taken with the close-up having a ruler and gray scale held parallel to the central point of interest.

It is common that victims will have injuries on several body parts. The presence or absence of trauma to a body part will assist investigators in corroborating the history provided by the victim or any witnesses. For example, the victim's hands and arms may reveal bruises, abrasions, or musculoskeletal trauma incurred during the struggle with the assailant. These wounds are termed *defensive injuries* since they are the result of fending off blows or self-protective maneuvers during the attack (see Figures 2.31 and 2.71).

The nature and extent of injuries often determine the sequence of recording of photographic images. However, many experts suggest that you start with orientation photos; then proceed to the extremities, chest, and abdomen prior to obtaining images of genital regions.

Drape parts of the body that are not essential to the photography of any specific area or detail. Avoid extraneous exposure of breasts, buttocks, the perineum, or genitalia. When the situation dictates the recording of details in these sensitive areas, careful draping of adjacent parts should be accomplished. A longer focal length macro lens that permits the photographer to remain well out of the patient's personal space should be used if available.

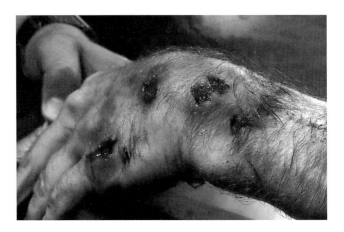

Figure 2.71 Defensive injuries. Courtesy Dr. Patrick Besant-Matthews. Used with permission.

Although most injuries can be successfully documented using any standard 35-mm or digital camera, there are some indications for special equipment. A common adjunct is an electronic flash unit to augment natural light. The use of flash should be reserved for those items that are visible to the naked eye. Care should be taken against overexposure that can actually make injuries less obvious due to reflection from the patient's skin. Holding the flash unit at an angle to the skin may help in guarding against this common problem. A ruler and color and gray scale should always be included as a reference. The photographer must position size and color reference tools at a nonreflective angle to the camera and flash unit.

Specific techniques using alternate light sources may be employed by those who perform evidentiary examinations for sexual assault and child or elder abuse cases. An important alternate light source for forensic work is reflective ultraviolet photography that will reveal bite marks or bruises that are not apparent with white-light illumination. Long-wave ultraviolet penetrates the skin, highlighting soft tissue trauma. Special filters must be used on the camera lens in order to block light other than the UV source. Light-emitting-diode (LED) technology is another useful, inexpensive, and portable tool for use in forensic cases. Operators must understand various wave lengths and the applications in certain settings and within color ranges. For instance, blue color ranges highlight body fluids such as semen, saliva, or urine; purple enhances bruises and abrasions. When using any alternate light sources, the photographer should have special training and experience in using the equipment properly.

Figure 2.72 Camera mounted on colposcope. Courtesy Janet Barber. Used with permission.

In the hospital setting, a colposcope-mounted camera may be used to elicit highly magnified images of soft tissue trauma associated with sexual assault injuries (see Figure 2.72).

Tips for Photographing Anatomical Features

When photographing rounded anatomical areas such as the face, breast, or buttocks, several angles and perspectives should be obtained. If the photographer knows the approximate position of the subject at the time the injury was inflicted, this position should be duplicated and the injury photographed to appreciate the relative mechanics associated with the wound. For example, if a female subject was standing upright when her breast was bitten by an assailant who was standing in front of her on the left side, the bite mark should be photographed with the subject standing upright and the photographer oriented to duplicate the relative position of the attacker.

Clothing with blood stains, cuts, tears, or holes should be photographed too, giving special attention to preserving any possible trace evidence such as sand, rocks, glass shards, hairs, or fibers. When all wounds have been photographed, the patient's ID should be photographed once again before the memory card is removed from the camera. Images should be duplicated and stored per local policies and procedures. The original should be properly labeled and sealed as evidence. No images, even those that are fuzzy or out of focus, should be deleted or altered in any way. All original images that have been recorded must be submitted as evidence.

Digital Evidence

Forensic science has discovered new challenges as well as vast opportunities associated with the digital age. Computers, voice and video recorders, surveillance cameras, and many wireless devices used in homes, public spaces, and within healthcare venues provide new sources of event documentation as well as forensic evidence for law enforcement and the legal systems. Forensic personnel must be aware of these important sources of information and take safeguards to ensure that digital information is not disregarded, lost, or destroyed during routine encounters with potential crime scenes, victims, or suspected offenders.

What Is Digital Evidence and Cyberforensics?

Digital evidence is any information that is stored or transmitted by digital devices, which can be recovered and analyzed to determine facts and circumstances related to a forensic event. The detection, analysis, and reporting related to evidence found in the physical and virtual memory of any computer is termed *computer forensics* or *cyberforensics*.

Digital evidence can be obtained from:

- Computers
- Scanners
- Compact flash cards
- Compact discs
- Digital audio and video recorders including telephone answering machines
- Digital cameras
- Cell phones
- Digital fax machines
- Personal digital assistants (PDAs)
- Other handheld digital devices

What Is the Unique Forensic Value of Digital Evidence?

Information contained in digital devices can be used to:

- Establish identities of individuals
- Reconstruct the sequence of events
- Determine movement and location of forensic subjects
- Provide precise documentation of details by digital photographs or recordings

- Confirm the relationships between people and physical locations at exact points in time
- Appreciate the typical behaviors of victims or offenders

Forensic investigations often use digital evidence to supplement physical evidence to draw conclusions about a forensic event. Three specific phases are essential: (1) data acquisition, (2) analysis. and (3) presentation. Every organization (i.e., business, industry, healthcare organization, military, or other governmental agency) has a unique set of standards and organization policies in regard to computer usage and recovery of any stored data.

- Data acquisition involves saving, copying, or producing an exact image of allocated and unallocated space on the hard disk
- Analysis of the data examines files and directories as well as recovering deleted content
- Presentation of findings, with a distinct focus on policy and law

Governmental and other public agencies as well as various types of business enterprises have their unique policies and procedures regarding acquisition and use of digital evidence. Hospitals and healthcare agencies have detailed protocols as well, but employees are typically not aware of the specific provisions. However, when legal actions occur, healthcare workers may be surprised that their personal devices may be confiscated in order to recover or corroborate important evidentiary details.

The United States has distinct requirements for lawfully intercepting communications or obtaining information from voicemail, e-mail, and text messages. Search or eavesdropping warrants or consents of a customer may be required in certain instances. The legal basis for capturing digital data related to criminal forensic investigations associated with drug trafficking, manslaughter and murder, armed robbery, or other serious offenses is *probable cause* as defined by the Fifth Amendment of the U.S. Constitution.

Healthcare workers must be aware that some situations mandate an immediate release of digital data. For example, the content and caller ID associated with emergency 911 calls, ambulance dispatch records, emergency department switchboards, and public agencies such as police and fire departments must be made available in real time. Crisis lines for suicide prevention or other personal issues may not have the same provisions, however (Hoffman and Terplan, 2006, 345). Forensic investigations are often able to use data from such resources to reconstruct events associated with either victims or suspects. However, they must have a clear understanding of the policies and procedures that pertain to the acquisition and use of such data.

Digital data sources that may prove to be valuable in forensic investigations include:

- Archived CD-ROMs or other recordings of 911 calls
- Personal cell phones or other handheld devices used by victims or suspects
- Recordings of communications between emergency personnel and hospital emergency department staff
- Surveillance camera recordings
- Computerized medical records
- Digital recordings captured by monitors or other medical devices
- PDAs (personal digital assistants) of staff members
- Tracking records of personnel, equipment, supplies, or pharmaceuticals

Although the Health Insurance Portability and Accountability Act (HIPAA) protects patient privacy, the intent of the act is to prevent unauthorized and accidental disclosures of medical information during transmission within and among medical facilities, healthcare workers or providers, and third-party payers (Frank-Stromborg, 2006).

Implications for Healthcare and Forensic Personnel

The use of both wired and wireless digital devices poses certain security risks for both patients and their caregivers. Some of the devices are strategically placed and designed to record sequence of events, personnel actions, physiological information, and communication transactions. Much of the record keeping within and among such devices occurs seamlessly in the care environment, often even without the knowledge of the caregiver or the patient. Events are timed, tracked, and recorded, and personnel interactions with such devices are also documented on embedded software that in turn is linked to the hospital information system. Automated drug dispensing units precisely record who accessed the locks and removed various items at a specific time. Some are capable of sending automatic messages to the pharmacy to control inventory or to a patient's chart to generate charges or to record administration. Hospital beds may be equipped with smart technology that periodically downloads its status. For example, beds contain software that records whether or not the fall-prevention system is armed, the height of the bed and any articulations such as the degree of head elevation, which side rails are up or down, and if the brakes are properly set. This type of information assists nurses in routine safety monitoring and becomes quite valuable in determining a root cause if a patient suffers from a bed-related fall.

Hospital personnel may be surprised about the number of ways that their actions are watched and recorded. It is imperative to understand that events are also time stamped regularly and if written documentation is used, care must be taken to ensure that serious time discrepancies are not introduced. If a cardiac monitor detects ventricular fibrillation at 3:09 p.m. and the nurse uses her personal wristwatch to determine when the code was called, even a few minutes of discrepancy between the two times could create doubt about the staff responsiveness to the emergency event. Assume that the in-house physician's cell phone records indicate that he received the call at 3:14 p.m. but the unit clerk logs that the call was placed at 3:11 p.m. according to a wall clock at the nurses' station. If the patient does not survive the resuscitation attempt, and a wrongful death suit is brought against the hospital several months later, considerable debate will be required to decipher the sequence of events and the precise timing. Hospital personnel should always use digital time references from a computer or cell phone rather than rely on personal timepieces such as a wristwatch. Elapsed time clocks are desirable to precisely record the overall length of an event such as a cardiac resuscitation.

When lawsuits allege negligence, malpractice, or medical device malfunction or failure, digital documentation can prove extremely valuable to confirm the actions of caregivers, especially when there is suspected tampering of the written record. A late entry into a record is viewed as a red flag by those who audit or review medical records. When it is recognized that an important data point was not added to the record in a timely manner, it may be documented later, along with the annotation "late entry." The date and time should also be recorded. It is unacceptable to insert a supplemental page expressly for such late entries. Even brief notes should not be added to the margins of chart pages or inserted between existing lines of an earlier note. All late entries are certain to be scrutinized if there are legal issues about the patient's care. This is especially true if the late entry pertains to a time period immediately before or after an adverse event.

Occasionally, entries are made into another patient's record by mistake. These errors should never be corrected by obliterating the mistaken entry by using white-out fluid or a black marker. A single line should be drawn through the entry, annotating, "mistaken entry" on the line. The date, time, and a signature should accompany the correction.

Tampering and Spoliation of Records

Tampering is an intentional act to secretly corrupt or alter the record in an attempt to influence its value as an item of evidence.

Spoliation is an intentional act to avoid discovery of wrongdoing by altering records. It can take many forms, ranging from subtly changing entries to completely destroying an entire medical record file. Acts of spoliation usually are committed by those implicated with some sort of involvement or those who are guilt ridden about an untoward medical event such as a medication error, patient fall, or other devastating event that is likely to precipitate a legal action.

Spoliation actions include:

- Altering record entries
- Destroying records
- The design of fabricated entries that support one's guilt or innocence
- Threatening witnesses not to testify
- Willful destruction of property that could be used to support one's cause, defense, or a specific claim
- Recording procedures or treatments that were not completed in order to collect nonearned monies from third-party payers or to gain other financial advantages

For example, the trashing of EKG tracings after a botched code, the destruction of a drug ampule after a medication error, or the willful failure to preserve material evidence such as a faulty infusion pump could all be termed *spoliation acts* in terms of legal proceedings.

If an attorney can show that there is evidence that has been lost, destroyed, or altered, a significant advantage is gained in the legal proceedings.

Safety and Security Concerns

Medical records should be maintained as *secure documents* and kept in a location where they cannot be altered. With the advent of electronic records, new challenges exist in regard to prevention of unauthorized tampering or spoliation. Even though passwords are issued, it is easy to become complacent about them when working among cohorts who are known and trusted. However, passwords should never be revealed or shared with coworkers. When password access has been used to open a record, the file should not be left open if unattended. If the computer must be abandoned, even for a minute or two, the healthcare worker should log off promptly to prevent others from making entries. Since electronic records have automated time and date stamps, it is fairly easy to discern when others change or delete data. Detectable electronic footprints remain that permit expert investigators to reconstruct all documentation transactions. Additional protection can be realized by the use of an infrared radio frequency (IR/RF) system to link caregivers to a recording device. An employee badge serves as an electronic key to access the computers or certain medical devices. It is imperative that such badges be carefully controlled to ensure that they do not become lost, misplaced, or stolen, which might lead to misuse of the key. Careless misplacement of the badge could permit other employees to steal your identity in order to gain access to controlled narcotics, medical records, or off-limit locations within the hospital.

Most hospitals have automated record back-up systems and have safeguards such as encryption to prevent computer hacking and unlawful access of individuals who might be intent on medical-record tampering. The use of a laptop at the bedside poses vulnerabilities for the safe maintenance of sensitive patient information. For example, a bedside laptop can be accessed and information downloaded using a simple flash drive. Such actions could compromise the security of both medical and financial information of many patients. The use of personal cell phones or other handheld devices that can record images as well as voice communications can compromise patient privacy and serve as media for surreptitiously recording information from monitor screens, labels from infusion bags, the appearance of dressings or wounds, contents of drainage bags, and even the actions or photos of caregivers.

Reporting

Any individual who believes that they have witnessed electronic medical-record tampering must report it at once to computer security personnel. Other nondigital tampering acts should be promptly brought to the attention of supervisory personnel.

Appendix 2.1: Adult Male Body Diagram

Name: _____ Case #: _____ Date: _____

Courtesy Dr. Patrick Besant-Matthews. Used with permission.

Appendix 2.2: Adult Male Body Diagram, Side View

Name: _____ Case #: _____ Date: _____

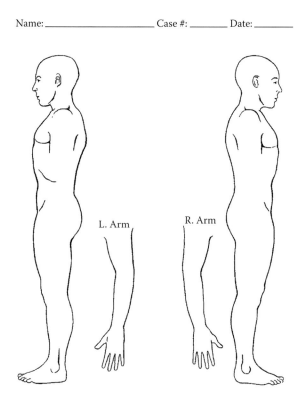

Courtesy Dr. Patrick Besant-Matthews. Used with permission.

Appendix 2.3: Adult Female Body Diagram

Name: _____ Case #: _____ Date: _____

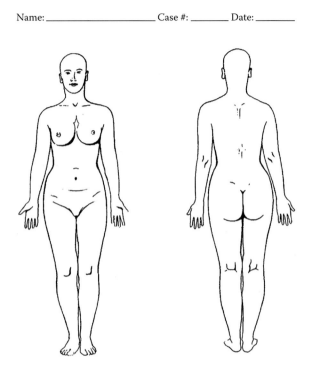

Courtesy Dr. Patrick Besant-Matthews. Used with permission.

Appendix 2.4: Adult Female Body Diagram, Side View

Name:_____ Case #:_____ Date:_____

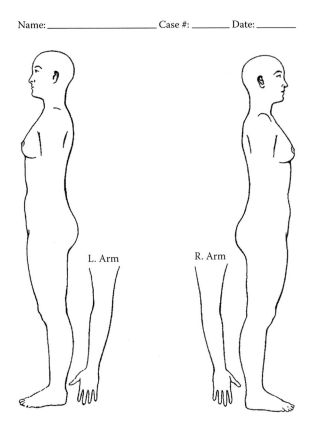

L. Arm   R. Arm

Courtesy Dr. Patrick Besant-Matthews. Used with permission.

Appendix 2.5: Head and Neck

Name: _____ Case #: _____ Date: _____

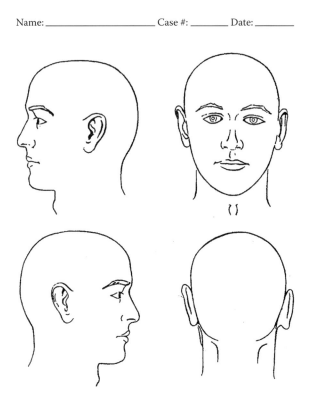

Courtesy Dr. Patrick Besant-Matthews. Used with permission.

Appendix 2.6: Hands

Name:_____ Case #:_____ Date:_____

Courtesy Dr. Patrick Besant-Matthews. Used with permission.

Appendix 2.7: Feet—Toes and Bottom

Name:_____Case #:_____ Date:____

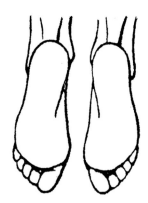

Courtesy Dr. Patrick Besant-Matthews. Used with permission.

Appendix 2.8: Feet—Side View

Name: _____ Case #: _____ Date: _____

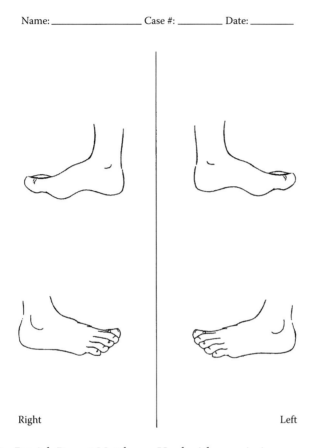

Right   Left

Courtesy Dr. Patrick Besant-Matthews. Used with permission.

Appendix 2.9: Hands, Feet, Head, Eyes, Ears

Name: _____ Case #: _____ Date: _____

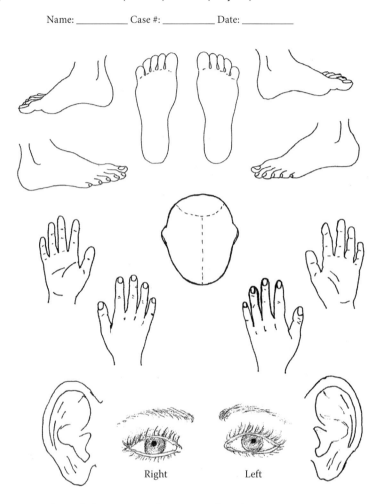

Right   Left

Courtesy Dr. Patrick Besant-Matthews. Used with permission.

Appendix 2.10: Female Genitalia

Name: _____ Case #: _____ Date: _____

External Female Genitalia

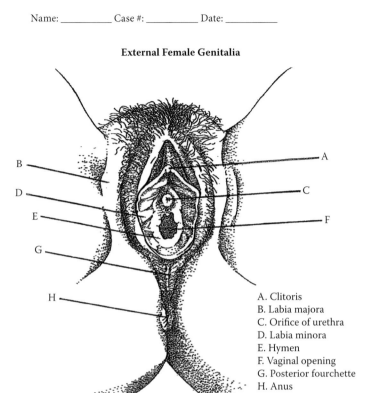

A. Clitoris
B. Labia majora
C. Orifice of urethra
D. Labia minora
E. Hymen
F. Vaginal opening
G. Posterior fourchette
H. Anus

Courtesy Dr. Patrick Besant-Matthews. Used with permission.

Appendix 2.11: Male Genitalia

Name: _____ Case #: _____ Date: _____

External Male Genitalia

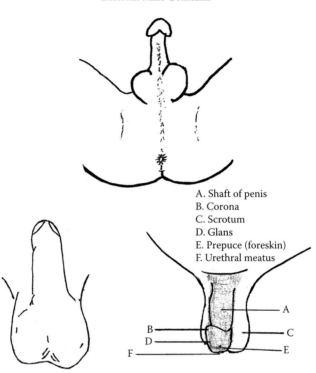

A. Shaft of penis
B. Corona
C. Scrotum
D. Glans
E. Prepuce (foreskin)
F. Urethral meatus

Appendix 2.12: Child Body Diagram

Name: _____ Case #: _____ Date: _____

Child Body Drawing

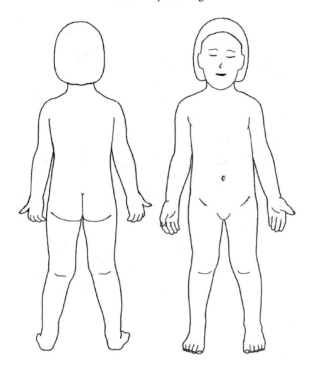

Courtesy Dr. Patrick Besant-Matthews. Used with permission.

Appendix 2.13: Infant Body Diagram

Name:_____ Case #:_____ Date:_____

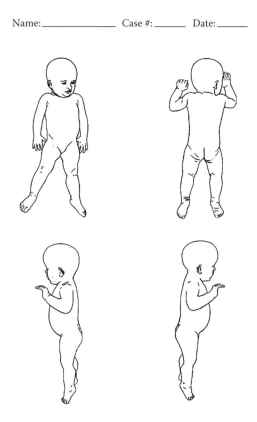

Courtesy Dr. Patrick Besant-Matthews. Used with permission.

Appendix 2.14: Consent to Photograph Forms

Consent to Photograph

I, _____ consent to being photographed by _____ and its staff, employees or other authorized persons while under the care of this facility or agency.

I have been informed and understand that any photographs taken may become a part of my permanent medical record and may be subject to subpoena.

I understand photographs in my record may be released if they are requested by a person authorized to obtain my medical record. If I do not want photographs released with my medical record, I must specifically exclude them in any authorizations I sign.

I (do ____) (do not ____) consent to use of these photographs for educational purposes.

I (do ____) (do not ____) consent to publication of these photographs for educational purposes.

_____

Signature

_____

Street Address

_____

City   State   Zip Code

Dated: _____

_____

Witness

Deposition of Photographs: To be completed by Staff

_____ (Number of photographs taken)

_____ Photographs were placed in a sealed envelope marked with patient's name and medical record number and were sent to medical records.

_____ Photographs were stored digitally according to policy.

Consent for Educational, Research, Public Relations, or Publishing of Photography

_____   _____
Name   Date

_____ _____   _____
Date of Birth   Age   Sex

_____   _____
Name of Agency, Individual or Institution   Case or File Number

_____
Address of Agency or Institution

I understand these materials may be included in my permanent medical record or the permanent records of the above agency and may be subject to subpoena.

I also authorize the above agency, individual or institution to interview, photograph or make other visual or audio recordings of me. I also authorize the above agency, individual or institution to use or permit others to use, the interview, photographs, or recordings for staff education, publication, public relations, and/or any other purpose or manner as they may determine to be appropriate. This permission is given of my own free will and without coercion subject to the following restrictions and/or limitations:

_____

_____

I acknowledge that I have voluntarily given this authorization without expectation of payment or other compensation, whether now or in the future. I hold the above-named institution harmless from and against any claim for compensation or harm resulting from the activities authorized by this agreement and extend this agreement to my family, heirs and assigns.

_____   _____
Patient or legal representative (signature) Date and time

_____   _____
Printed name of patient or legal representative Witness or translator

Appendix 2.15: Some Commonly Used ICD-9 and CPT Codes

Diagnostic ICD-9 codes (Hart, Stegman, and Ford, 2009) for Adult Maltreatment and Abuse: (995.8)

- Physically abused person, battered person, spouse or woman
- Adult emotional/psychological abuse
- Adult sexual abuse
- Adult neglect (nutritional)
- Other adult abuse and neglect (multiple forms)

CPT codes are used in conjunction with ICD-9 diagnostic codes. Codes are not specific for domestic violence or sexual assault. However, the following codes are examples of ones that might be used in the evaluation and treatment of a victim of interpersonal violence (Beebe, 2009).

- 99303   Complex evaluation and management: comprehensive history, comprehensive examination, medical decision-making of moderate to high complexity, counseling and/or coordination of care with other providers or agencies
- 99361-2 Team conference
- 99374-5 Care plan oversight
- 99381   Preventive medicine services
- 99401   Preventive medicine counseling

References

Alaska protocol. Hypothermia prevention, recognition and treatment. Articles, protocols and research on life-saving skills. http://www.hypothermia.org/protocol.htm/rev1/2005 accessed August 28, 2010.

Beebe, M., et al. Nov. 2008. Current Procedural Terminology. 2009 CPT Standard Edition: AMA.

Emedicine Health: http://www.emedicinehealth.com/burn_percentage_in_adults_rule_of_nines/article_em.htm, 8/24/2010

Frank-Stromborg, M., K. Burns, and D. B. Sisneros. 2006. Health insurance portability and accountability act (HIPAA) in P. Iyer, B. J. Levin and M. A. Shea (eds.) *Medical Legal Aspects of Medical Records* (Chapter 11). Tucson, AZ: Lawyers and Judges Publishing Co.

Hart, A. C., M.S. Stegman, and B. Ford. 2008. 2009 ICD-9-CM, *Expert for Physicians*, Vol. 1 and 2. (6th Ed.) Salt Lake City, UT: INGENIX.

Hoffman, P. and K. Terplan. 2006. *Intelligence support systems: Technology for lawful intercepts*. Boca Raton, FL: Auerbach Publications. 345.

JAMA protocol. *Journal of the American Medical Association (JAMA)*. American Heart Association, Emergency cardiac care committee and subcommittees. October 28, 1992. Guidelines for cardiopulmonary resuscitation and emergency cardiac care. http://www.hypothermia.org/jama.htm. Accessed 8/28/2010.

James, J. H. and Nordby, J. J. 2009. *Forensic science* (3rd ed.) Boca Raton, FL: CRC Press, Taylor & Francis 55-56.

Joint Commission on Accreditation of Healthcare Organizations (JCAHO or "The Joint Commission"). 1995. *AMH Accreditation Manual for Hospitals*. Oakbrook Terrace, IL.

Lanros, N. E. and J. M. Barber. 1997. *Emergency nursing* (4th Ed.). Stamford, CT: Appleton and Lange. 510-511, 519-524.

Olshaker, J. S., M. C. Jackson, and W. S. Smock. 2007. *Forensic emergency medicine* (2nd Ed.). Philadelphia: Lippincott, Williams & Wilkins: 161.

Stewart, C. E. 1990. *Environmental emergencies*. Baltimore: Williams & Wilkins. 58-59, 290-291.

U.S. Coast Guard Protocol. Steinman, A. M. Hypothermia, drowning and coldwater survival. PVSS Conference. http://www.uuscg.mil/pvs/docs/coldwater1.pdf. (Accessed September 9, 2010).The protocol can also be accessed from http://www.uuscg.mil/hq//cgl/cgll2/docs/pdf/SAR_CPR_protocols.pdf.

Wagner, S. A. 2009. *Death scene investigation: A field guide*. Boca Raton, FL: CRC Press, Taylor & Francis, 101-104.

Witter, Robert. January 29, 2010. Unpublished presentation given to La Cueva Volunteer Fire Department, Jemez Springs, NM.

Evidence

3

Evidence is defined as the means by which an alleged matter or fact is established or disproved.

Legal evidence is that which is admissible in a court of law under the *rules of evidence* to prove or disprove a claim.

Widely accepted among the scientific community, Locard's *principle of interchange* states that a person visiting a particular environment will leave traces of his or her presence in that environment, and will carry traces of the environment away when he or she leaves, thus making an *exchange*.

For legal purposes, evidence may be information or a physical item admitted to the court for the purpose of determining the truth of a statement or fact under consideration. In cases of interpersonal violence, the injured individual may be the only witness, and his or her body may be the actual crime scene. As such, it is the source of many types of evidence that needs to be collected and preserved. If the victim fought back, the perpetrator may retain evidence of having been scratched or bitten by the victim. Trace evidence such as hair, makeup, and fibers from carpets or clothing may help link the victim to the perpetrator.

The Centers for Disease Control and Prevention (CDC) estimates that there were over 4 million visits to emergency services providers for injury-related problems in 2000 (Assid, 2005). Nurses, emergency department physicians, primary healthcare providers, and first responders are in an ideal position to observe and document these injuries.

The critical first step nurses must take is to develop a forensic antenna. By recognizing those who are victims within our patient population, we help to ensure that objective usable information is available to those making critical medical and legal decisions. The first step in this process begins by recognizing existing evidence in the presenting scenario. Once the injured party has been identified as having possible forensic issues, the healthcare provider must be knowledgeable regarding their role in collecting the evidence, documenting the collection process, and preserving the integrity of that evidence so it can be used in criminal or civil proceedings.

Healthcare Role in Evidence Collection

The Joint Commission on the Accreditation of Healthcare Organizations (JCAHO), or the Joint Commission (JC), requires accredited healthcare institutions to establish and implement a program to address victims of violence. Written policies and procedures are required, and all personnel must be trained to detect victims of violence, abuse, and neglect (JCAHO, 2009).* The Joint Commission further emphasizes that personnel must know how to preserve vital evidence and to take an active role in reporting their suspicions or findings.

Healthcare personnel can no longer assume a passive role in these cases without risking repercussions. Practicing healthcare professionals can be prosecuted for failure to cooperate with law enforcement. State boards of nursing also confirm the position that its members have a legal, moral, and ethical obligation to act in the best interests of the public. This includes gathering and preserving evidence.

In the clinical setting, it is important to recognize that the patient may be the crime scene. Each scene is unique, thus requiring those taking care of the victim to make complex decisions, often in a very short period of time. Evidence should be collected in a comprehensive, systematic manner without compromising delivery of medical care. To facilitate this process, evidence collection protocols should be developed in each and every healthcare setting. Although care of the traumatized patient is a multifaceted process, maintaining an organized environment allows those involved to think, avoid rushing, and proceed in a rational manner so that possible evidence is not overlooked. Evidence that is ignored, unnoticed, or not documented at this point is simply lost forever. It cannot be re-created later. In addition to the fact that all evidence may be important in proving the facts,† absent evidence leaves the prosecution with little to support their case. It also opens the door for the defense to raise doubts as to whether the "loss" may have been intentional.

Healthcare Providers

In order to effectively carry out these responsibilities, nurses must have education and training in evidentiary procedures including collection of physical and trace evidence, maintaining a chain of custody, and being

---

* Standards are general and must be implemented by each agency in policy and demonstration of practice.
† Many cases are based on the cumulative effect of overwhelming circumstantial evidence rather than items of direct evidence (crime.about.com, 2010).

able to demonstrate competency in written and photo documentation. Initial training as well as refresher courses should be available on a regular basis.

Regardless of the clinical setting, evidence should be collected based on recognized norms. Healthcare providers must understand the concepts, principles, and techniques for evidence collection, and realize that various jurisdictions may have differing standards or procedural requirements. Evidence needs to be handled according to local jurisdictional policies in order to hold up in court later on. Local, state, and federal laws differ in many areas. The healthcare provider needs to know how to access those laws and be able to recognize the nuances of each in order to properly manage evidence collection in their jurisdiction.

Understanding what resources are available in your particular community is invaluable, not only in the planning of referrals, but also to ensure that vital evidence is preserved to support any subsequent legal proceedings.

The following is a partial list of situations where the nurse might be required to collect forensic evidence:

- Trauma resulting from motor vehicle accidents (including automobiles, motorcycles, planes, boats, trains, or any other motorized apparatus)
- Any motor vehicle versus pedestrian trauma
- Attempted suicide or homicide
- Injuries involving firearms or other weapons
- Child or adult victims of abuse, neglect, or sexual assault
- Work-related accidents
- Accidents resulting from fires, falls, electrocution
- Poisoning, overdoses, or illegal drug use
- Trauma in unidentified victims or suspects
- Trauma of suspicious origin
- Injuries resulting from improperly used or malfunctioning equipment
- Injuries or illness from work-related incidents, environmental hazards, or pollution
- Anyone in police custody
- Sudden, unexpected, or unattended death
- Injuries resulting from suspected medical malpractice

Categories of Evidence

Evidence generally falls into two major categories: *Direct evidence* and *indirect evidence*. *Direct physical evidence* is that which proves a fact without inference or presumption. It is also known as *positive evidence* because it establishes a solid connection directly tying one thing to another. It is the uniqueness of a particular item that guarantees only one person can be tied to that particular piece of evidence. Examples of positive evidence are latent fingerprints and DNA.

Indirect physical evidence is that which establishes circumstances from which one can infer facts at issue. Indirect evidence, more commonly thought of as circumstantial evidence, is considered to be indirect because it implies that something occurred but does not directly prove it.

Circumstantial evidence can be one or more facts from which another fact can be deduced, or it can be a chain of events indicating that something did or did not occur (crime.about.com, 2010). The connection between the fact and the inference must be strong enough to be probable. In other words, there is a reasonable surety that the inference can be made.

Types of Evidence

Physical Evidence

Physical evidence consists of tangible objects that may help prove or disprove a statement at issue—any matter, material, or condition that may be used to determine the facts in a given situation. Physical evidence includes: clothing, biological evidence, trace evidence, missiles, projectiles, and foreign objects. Examples of physical evidence are blood, saliva, seminal fluid, urine, drugs, bite marks, clothing, footwear, glass, pollen, explosives, firearms, insects, and tire track impressions.

Trace Evidence

Trace evidence is physical evidence that is found at a scene or on a patient's body in small but measurable amounts. With the exception of DNA, most trace evidence is circumstantial. Examples of trace evidence include hairs, fibers, glass, wood, metal, paint chips, soil, botanical materials, biological specimens (saliva, semen, etc.), and residue from firearms, explosives, or volatile hydrocarbons (e.g., accelerants used to start a fire). Traces of gunpowder may be found on either the victim or perpetrator's hands.

When sexual assault is alleged, the perpetrator's pubic hair provides a link between the victim and the perpetrator. Taken alone, it does not prove the allegation. In concert with other evidence, however, it may prove that the sexual assault was, indeed, perpetrated by a certain individual. Likewise, dirt, paint chips, or blunt force injuries may link the victim to a scene or weapon. Such evidence may also lead investigators to discover the scene location or the object used as a weapon. For instance, a *patterned injury* of a belt buckle, may lead to the actual belt used and possibly to its owner.

Verbal Evidence

Verbal evidence consists of statements given at a deposition or a trial that tend to support or refute, negate, or contradict a fact at issue.

The following examples illustrate the differences in types of verbal evidence.

- *Direct evidence* consists of eyewitnesses and witness statements that directly tie one thing to another: "I looked out the window at midnight and it was snowing."
- *Circumstantial evidence* establishes a fact from which an inference can be drawn: "There was no snow on the ground when I went to bed. When I woke in the morning, there was snow on the ground."
- *Hearsay* is a statement referring to something said by someone other than yourself. As a general rule, medical records are considered hearsay (Quinn, 2005):[*,†] "Jim said that it was snowing last night."

It is important that the victim's and perpetrator's statements be recorded exactly as stated. Tape recording is acceptable, but often impractical, and a written transcript eventually needs to be made. Regardless of the initial method used, the patient or victim's words must be recorded precisely using quotation marks. *Statements must not be paraphrased.* If the speaker is talking too rapidly, have him or her slow down or repeat so you can record the wording correctly.

Questions to Ask

- **Who**? Was the assailant known to the victim? Describe the physical characteristics of the perpetrator (height, weight, ethnicity, clothing, smells, distinctive voice, etc.).
- **How many** perpetrators were there?
- **Were there any witnesses**?
- **What happened**? Describe the scenario in precise detail.
- **When**? Document the exact time, if known. Describe what activities were going on immediately prior to the incident.
- **Where**? Was the victim transported from the place of initial attack? How was the victim transported and by whom? How far?

---

[*] Medical records generally fall into the hearsay category. Notations in the medical record are recordings of second-hand information and are "out of court" declarations.

[†] Hearsay evidence is not admissible in court unless it meets certain criteria, which allows exception to the hearsay rule. The hearsay rule generally prohibits a person from providing testimony because the information is second-hand. The rules of evidence (formal court rules) favor testimony based on a person's own observation or knowledge. The subject of hearsay is complicated.

- **How was initial contact made?** What methods did the perpetrator(s) use to gain control of the victim?
- **Were weapons used?** If so, describe the characteristics of the weapon.
- **What did the perpetrator or perpetrators say?**
- **Were there any unusual circumstances?** (e.g., "It was thundering and raining hard. It was difficult to see or hear clearly.")
- In addition to what the victim saw, did the victim notice any distinctive sounds, smells, etc.?
- Listen to *how* questions are answered as well as to the substance of the answers.
- Are the injuries consistent with the explanation?
- Does the story change with time and by how much?

Demonstrative Evidence

Demonstrative evidence is the visual materials presented during legal proceedings. Examples are photographs, drawings, sketches, plaster casts or molds, scale models, maps, charts, body diagrams, police composite drawings, mug shots, computer reconstructions or animations, and scientific tests or experiments. Demonstrative evidence is intended to amplify or explain other testimony.

Testimonial Evidence

Testimonial evidence is evidence given in writing (such as in answers to interrogatories), or verbally (court testimony), or any other way that expresses a person's thoughts or observations. It may be obtained through inquiries, interviews, interrogatories, or questionnaires. A deposition is testimonial evidence.

Digital or Electronic Evidence

Digital or electronic evidence is any information stored or transmitted in digital form that may be used in a court of law. Digital evidence might include computer documents or data entry, social networking pages, intimidating or threatening e-mails, and photographs or video images captured on cell phones, surveillance cameras, or handheld personal devices. Other types of digital materials may be used as evidence in copyright proceedings, identity theft or other cyber crimes, unlawful disclosure of information, or proceedings regarding the ownership of intellectual property. Global positioning systems also may be used to track or locate persons or vehicles of interest in a crime.

Behavioral Evidence

Behavioral evidence can be defined as any acts or omission of acts that demonstrate a general or specific pattern of behavior or indicate a general or specific plan, objective, or purpose. The inclusion of behavioral evidence implies that cumulative behavior should be identified and documented so it can be compared to normal behavior, and become meaningful in light of the overall assessment of a patient or forensic situation as a whole.

Basic Principles of Evidence Collection

The patient's medical needs and safety are always the first priority. The process of collecting evidence need not interfere with lifesaving measures. Members of the healthcare community can become competent to identify, preserve, and collect evidence without compromising care of the patient. Proper collection of evidence is a systematic, comprehensive, and scientifically based process, beginning with recognition of potential evidence (both physical and verbal). It is an objective, logically driven endeavor.

Use knowledge and experience to extend your forensic antenna. In the nurse's mind, the possibility of forensic issues needs to be addressed immediately and ruled out or pursued from the moment the patient is first encountered. This means approaching the victim, offender, or witness, and evaluating their appearance and/or statements made with an index of suspicion. Because treatment or surgical intervention destroys evidence, initial photography and direct quotations from the patient should be done from the outset. Forensic techniques can be incorporated into your existing interview and treatment routine. If you are unable to photograph, use body diagrams or make a sketch. Initial notes should include direct quotes, incorporating them appropriately when preparing your final documentation.

Appropriate handling of evidence facilitates the prosecution of offenders; inadequate, sloppy evidence collection and preservation leads to its downfall. Likewise, memories easily deteriorate and the opportunity to capture statements made at the time the patient is first encountered will be lost.

Standard evidence collection procedures includes recording the person(s) from whom it was obtained and to who whom it was given, including the date and time of transfer. In the clinical setting, this is done in the medical record as well as on the evidence tag and chain of custody form given to police. A copy of the completed evidence tag should be placed in the medical record. See specific details in the "Chain of Custody" section of this chapter.

Comprehensive documentation is needed for continuity of care and medical review. When a case has forensic implications, either the prosecution or the defense may subpoena notes and photographs, making content critical. A sketch of the scene* along with thorough and accurate notes will be helpful when preparing final documentation (see Chapter 2).

General Procedures

Personal Protective Equipment

To protect both the integrity of the evidence and the health and safety of personnel, all persons coming into contact with victims of crime and trauma should always wear clean, powderless gloves and change them often. Wash your hands before putting on gloves and rewash when gloves are removed. Always wear clean, powderless gloves when collecting items of evidence. Do not touch the victim/patient with your bare hands. Biological material may contain hazardous pathogens, including the HIV and hepatitis C viruses. Although manageable to some extent, they are not yet curable and their associated disease processes can be debilitating and deadly.

Change gloves between handling of biological evidence to avoid cross-contamination, especially if the gloves become contaminated, stained, soiled, or torn. Change gloves between packaging large items such as clothing and collecting small or trace evidence items.

To protect yourself from contamination, avoid touching any area where body fluids might exist. Be especially careful when handling wet items. Keep any contaminated surfaces, including gloved hands, away from your face to prevent contact with mucosal surfaces of the nose, mouth, and eyes.

Collection Basics

Trace evidence is fragile by its very nature. According to Locard's principle, it may be transferred from the patient to the caregiver and then to another person or item. Items of trace evidence such as sand, plant material, hair, and so on, are often loosely attached. This is why they transfer easily from their original source. Likewise, they can fall or brush off of the intermediate carrier—the victim or perpetrator. This is the underlying reason that linen as well as the victim's clothing must be packaged—it has the potential to carry trace evidence.

Package each item separately. Never put two items in the same container. Cross-contamination between items of evidence can be avoided by placing only one item in a package. Place each item in a new, unused paper bag or box. Do not allow one evidentiary specimen, particularly those containing

---

* A simple diagram will do. Mark the sketch as "not to scale."

biological specimens, to come into contact with another. Collect each stain or other biological evidence droplet separately.

Do not use plastic bags or containers to package evidentiary items. When moisture is trapped inside, it allows the formation of mold, mildew, and other biological processes that can deteriorate the specimen and render it unusable.

Handle items of evidence as little as possible and in a manner that minimizes deterioration (e.g., keep away from heat and chemicals).

Collect as much sample as possible from each source. Always take control samples if possible.

The size of the container should correspond to the size of the object collected. Make sure that the paper container is large enough to allow air circulation around the evidence item, but not so large that the specimen can be tossed around.

Do not place evidence samples on a contaminated surface. Thoroughly clean any surface before using it to package evidence. Avoid cross-contamination between each item collected.

Do not talk, cough, sneeze, or blow onto biological evidence. Do not lick the glue on envelopes to moisten them for sealing. Use a moistened cloth, sponge, or paper towel. Keep medical specimens separate from forensic specimens.

Store metal or glass items at room temperature and submit them to law enforcement or the crime lab as soon as possible.

Keep in mind that evidentiary materials include examination documentation and photographs. Proper recording of exam findings and preservation of evidentiary materials is critical for admissibility in a court of law.

Remember to follow jurisdictional procedures for documentation of examination findings, medical forensic history, drying, packaging, sealing, and labeling of evidence.

Drying

Air-drying is the preferred method of preserving damp or wet evidence samples. Dry each item separately, out of drafts and high-traffic areas.

If drying is impracticable or impossible, the items may be refrigerated or frozen, but should be turned over to law enforcement as soon as possible. Contact local law enforcement for advice on local protocols.

- Liquid blood samples should be refrigerated, but not frozen.
- To prevent any possibility of contamination, do not dry samples from one patient in the same area as those from another patient.
- *Never* expose damp or wet items to heat.
- Do not leave samples or evidence unattended while drying.
- *Never* place items of evidence that may contain DNA in a plastic bag or container.

Labeling

All items of evidence must be clearly identified and labeled and secured for later use in court.

Label each evidence bag or container with the following:

- Patient's full name
- Patient's age and date of birth
- Patient's ID or medical record number, or law enforcement case number
- Brief description of the source including the location from which it was collected
- Date and time of collection
- Initials or signature of person collecting and packaging the evidence. Write this across the edge of the tape onto the evidence container itself

To avoid damaging or altering small or delicate evidence, write on the envelope before placing items of evidence inside. Envelopes or small containers may need to be placed into a second appropriately sized bag or container. To avoid damaging or altering evidence, it is a good practice to prelabel all bags and containers before placing the evidence inside. Note the contents of each bag on the outside of the bag.

Prepare a chain of custody form for each item collected and attach it to the bag or container. Record in your notes the total number of bags released as well as the name and badge number of the person receiving items of evidence (detective, law enforcement officer, medical examiner's office representative, etc.) and the date and time you released the evidence. This will serve as a cross-reference to the chain of custody form in case it is misplaced or lost.

Sealing

Each bag must be sealed so its contents cannot be tampered with or removed and nothing can be added. Place tamperproof evidence tape across the edges to be sealed. Proper sealing is accomplished by securely taping the container, and initialing and dating the seal by writing over the tape onto the evidence container. Do this in at least two places. This practice makes it more difficult to remove and replace the tape exactly as it was placed initially. Completely seal all edges of the outer container so that attempted tampering could be detected.

Staples may enhance closure, but are not considered a secure seal. Tamperproof tape is the only secure method of ensuring a seal is intact. To ensure an envelope or container is tamperproof, all outer edges must be covered securely with tape (see Figure 3.1).

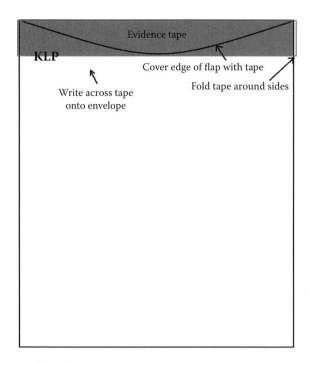

Figure 3.1 Sealed envelope.

Never seal gummed envelopes by licking; use a moistened paper towel, sponge, or clean cloth moistened with plain water. Sponge-tipped, reusable water-filled envelope moisteners are inexpensive and are available at office supply stores (see Figure 3.2).

Processing
Not all evidence that is collected is processed. There are times when processing of evidence may be unnecessary (the perpetrator pleads guilty) or it is simply not cost-effective. Nurses still have a responsibility to collect it.

Storage and Security
Documentation of the patient examination, including sketches and photographs, are considered evidentiary items. Follow jurisdictional policies regarding chain of custody and release of documentary evidence.

If the patient dies, the medical examiner or coroner may want to see the body before it is taken to the morgue. Contact their office before moving a deceased person.

Figure 3.2 Bottle moistener.

Chain of Custody

To retain its legal value, there must be a clear trail showing that the item presented in court is exactly the same as the one found at the scene. This is accomplished by a written recording of every person who had possession of the item of evidence from the time it is collected until the time it is presented in a court of law or other legal proceeding. This procedure is called the *chain of custody*.

The chain of custody is the knowledge or record of each person who has come into possession of a physical object from the time it was discovered until it is presented in court. To retain their evidentiary value, items of evidence must be unaltered and uncontaminated in any way. Any break, however small, can be lethal to the admissibility of that piece of evidence at trial.

Key Point:
The chain of custody is a procedure ensuring continuous accountability and validating that items of evidence are authentic, have been under constant surveillance, have been secured in a locked area, and have remained unaltered during handling and transfer.

The chain of custody begins with the person who first collected the evidence. It must be maintained throughout legal proceedings regardless of who originally obtained it. At trial, the custodian must be able to demonstrate that the item presented in court is the same as the one originally collected. He must be able to show that it has not been altered or tampered with except for examination by the criminologist. All persons handling items of evidence must protect it from damage, deterioration, or loss. Any evidence damaged, misplaced, or lost can be deemed inadmissible. Careless handling of evidence also reflects poorly on the professionalism of the responsible agency. The security and integrity of evidence is the responsibility of all persons who may identify, collect, package, store, transport, or examine evidentiary items. In the medical setting, initiating and maintaining the chain of custody is the responsibility of the licensed professional.

The chain of custody form includes a description of the item, the date and time it was collected, and the name and signature of the person who collected it. Each successive custodian must note their name, title, date, time, signature, and the agency they represent. Appendix 3.1 is provided as an example.

When all evidence has been collected, each bag or container should have a chain of custody form firmly affixed on the outside. Items should be numbered sequentially and a list should be made. A copy of this list should be attached to the nurse's notes or a corresponding note made directly into the progress notes in the medical record. The following is an example:

Bag 1 of 7   left shoe
Bag 2 of 7   left sock
Bag 3 of 7   right shoe
Bag 4 of 7   right sock
Bag 5 of 7   underwear
Bag 6 of 7   pants
Bag 7 of 7   shirt

The chain of custody form indicates all persons releasing and receiving custody of the item, and the date and time it was transferred. In the clinical setting, this is done in the medical record as well as on the evidence tag and chain of custody form that accompanies the item of evidence. A copy of the completed tag and/or a complete list of items of evidence should be placed in the medical record.

Transfers should be kept to a minimum. The number of individuals handling and transporting evidence should be kept to as few custodians as possible. This reduces the chances of compromising the chain, thus minimizing the opportunity to challenge the validity of the evidence at trial.

The patient should never be left alone with items of evidence—either potential or collected specimens. Patients, family members, or other bystanders should never be given responsibility for safeguarding or transporting physical items of evidence.

When items of evidence are transferred to law enforcement, a signed receipt should be obtained from the law enforcement representative and placed in the patient's medical record. In cases where evidence cannot be turned over to law enforcement in a timely manner, all items must be locked in a secure cabinet with only one key readily available. That key should be kept by the designated custodian, preferably the licensed professional, house supervisor, or security officer. If the key is turned over to another individual, a chain of custody documentation must occur. Forensic evidence should not be stored in a narcotic cabinet, unit refrigerator, or staff locker.

It is not necessary to maintain a chain of custody on medical specimens. Specimens collected for medical purposes should remain at the hospital for routine processing. Biological specimens (for example, blood for toxicology) taken for medical reasons are not treated as evidence. Generally speaking, specimens for law enforcement are taken separately. If law enforcement requires medical specimens, the patient's consent or a subpoena must be obtained before the specimens are released to law enforcement and the occurrence noted in the progress notes along with a copy of the subpoena or signed consent form. Check with your local district attorney regarding current law in your state.

Common Types of Physical Evidence in Medical Settings

Clothing

Staff should be prepared to treat all clothing as potential evidence. When the patient is the victim of a crime, trauma, or other medicolegal situation, collect all garments and personal items, including scarves, hats, shoes, undergarments, hair ornaments, jewelry, and so on.

Inspect clothing for damage, foreign material, or stains. Note any rips, tears, stretching, missing buttons, or other signs of damage. Also note any apparent biological substances such as blood, saliva, hair, seminal fluid, and/or other materials such as dirt, grass, or other debris, and collect as biological or trace evidence.

To avoid cross-contamination, wet or damp clothing should be air-dried before packaging.

Place each item in a clean, separate paper bag—one item per bag. Plastic containers are not suitable because damp materials will degrade biological materials (blood, body fluids, tissue, etc.), possibly rendering the item unusable for evidence recovery.

If you do not have secure facilities to dry clothing, place clean sheets of white paper between folds to protect from cross-contamination and package as above. If you think an article is wet enough to leak through a paper container, it can be placed on, *not in*, a plastic sheet. Notify law enforcement that you have wet items that need to be picked up and dried as soon as possible. Law enforcement should know the importance of drying, the proper procedures, and have evidence-drying cabinets to handle wet materials. The only exception to packaging wet clothing in paper is the presence of volatile or explosive chemicals (such as might be found in a suspected arson fire). These garments should be placed in airtight containers and law enforcement must be notified immediately.

Gently fold clothing inward, placing a clean piece of white paper against any stain that might come into contact with another part of the garment or the inner surface of the bag itself. Never shake clothing or handle it roughly.

Ask the patient if the clothing he or she is wearing now is the same clothing worn at the time of the injury. If the original clothing is not present, try to determine its location and notify law enforcement immediately.

If possible, have your patient remove his or her own clothing. If the patient can stand, place two clean sheets on the floor. The top sheet will be collected and bagged as evidence. The purpose of the top sheet is to catch any loose evidence that might fall from the clothing or body while disrobing. The purpose of the bottom sheet is to protect the top sheet from dirt and debris on the floor. Fold the sheet into a bindle to ensure that all foreign materials are contained inside (see Appendix 3.2). Hospital and prehospital linens should also be collected—one item per bag. This includes patient gowns, stretcher or gurney sheets, operating room table linen, and drapes.

Handle clothing gently to prevent loss of trace evidence and preserve signs of force (e.g., torn buttons, stretched fabric, etc.). *Never shake clothing.*

Regardless of who removes the clothing, *do not* cut through any disruption in the fabric. *Do not* cut through holes made by trauma. Instead, cut along seam lines or well outside the margins of rips, tears, holes, or other areas that might be of evidentiary value. If clothing was damaged or cut by healthcare personnel, document this in the nursing record.

Bullets and Other Projectiles

Handle bullets and other penetrating objects very gently. It is important that no additional markings are made on bullets or other penetrating objects. Use gloved hands if possible. Sterile or new disposable instruments should be used. If tweezers or forceps are required to extract foreign objects, make sure the jaws are smooth (no teeth) and padded to avoid marring the surface. Plastic or rubber instrument tips are available from the operating room (OR) (see Figure 2.37). If rubber tips are not available, use 2 × 2 gauze as alternative padding. Document the fact that padded forceps were used in the medical record.

To clean tools between specimens, wipe them under running water and thoroughly dry with a paper towel. Repeat this process twice before using the tool to manipulate or obtain another sample.

Air-dry the projectile and gently wrap it in a 2 × 2 gauze to protect it. Place it carefully into a sterile, sealable receptacle (e.g., urine specimen container). Do not drop a bullet or other missile into a metal basin. Make sure the item is dry before sealing and attaching evidence labels. Labeling the container is sufficient to maintain the item's identity and chain of custody.

If more than one missile is collected, use a separate container for each.

Foreign Objects

Foreign materials collected from the patient's clothing or body may be of significant forensic value. They can be compared to similar evidence collected from the scene or to an assailant. They may also provide information regarding the circumstances of an assault. Examples are hair, fibers, glass shards, paint chips, soil, sand, grass or other vegetation, or unknown materials.

Always air-dry vegetation or damp soil samples prior to packaging. Otherwise, mildew or mold growth may occur, potentially making the sample less valuable for evidentiary recovery.

If soil or other foreign materials are firmly attached to some object or garment, *do not remove*, but air-dry and place in a clean paper bag. Again, label with patient's name and date of birth (if known), collector's signature, date, time, and case number. If soil is present on the item (e.g., shoes), make a notation on the outside of the evidence container (such as, "Sand particles present on shoes").

Trace or Small Evidence

Trace evidence is physical evidence so small it can easily be overlooked by a healthcare provider or detective. Common examples are gunshot residue, pollen, paint residue, fingerprints, and chemical residues.

A cotton-tipped swab is probably the standard collection device for trace or small evidence items. Include the swab in the package with any fragment or small item of evidence before sealing.

A new method of collection allows trace evidence to be gathered using dry, durable Teflon surface wipes. The advantage of these wipes is that they are shred resistant making them a more effective method of gathering samples of trace evidence from rough or jagged surfaces. Dry sampling is also preferable in cases where the pieces of debris are too large to use solvent extraction methods effectively or to conduct microscopic investigations. Teflon surface wipes can be used for sampling explosives residue from other surfaces such as clothing, hands, and luggage.

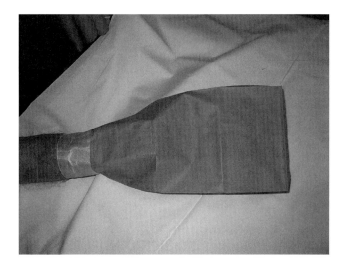

Figure 3.3 Bagged hand.

Gunshot Residue

Gunshot residue is extremely fragile evidence and should be collected as soon as possible—preferably within three hours of the shooting. In addition to hands, gunshot residue can be found on the clothing and skin of the victim, the person discharging the weapon, and on objects that were in the immediate vicinity of the weapon when it was fired. Each potential surface should be swabbed or collected, labeled with the exact site, placed in a separate container, and sealed, noting all of the above in the patient's chart.

If the patient was involved in a shooting, cover each hand with a brown paper bag and secure to the patient's skin with tape, mitten fashion (see Figure 3.3). Covering hands in this fashion preserves any gunshot residue that may be present. Liquor bottle bags are a good size for this purpose unless the patient has very large hands. Do not use white paper or plastic bags to cover hands.

Remove all clothing carefully as noted above. (Fold inward onto itself with clean paper between layers and place in a brown paper bag, labeling and sealing appropriately.)

Small or Loose Materials

Small or loose foreign materials such as fibers, soil, paint, splinters, and glass may be removed with clean forceps, using the sticky side of transparent tape, gently scraping the materials with a clean slide or the back of a scalpel knife, or cutting with a pair of clean scissors. Place in a labeled bindle (see Appendix 3.2, "Bindle Diagram and Instructions"). Place the bindle in a clean paper envelope and seal. Do not lick the gummed surface on the envelope. Use a dampened paper towel or gauze. Label the bindle with patient's name and date of birth (if known), collector's signature, date, time, and case number.

Large Loose Foreign Materials

Large loose foreign materials such as grass and hairs may be removed with clean forceps. Place in a clean paper bindle. Place in a clean paper envelope and package appropriately.

Fibers and Threads

Fibers and thread evidence are often found in torn or abraded areas of clothing or other materials. Examination of fibers and threads can be matched to the victim, perpetrator, clothing, bedding, carpet, vehicle, proving an exchange was made from one to the other. Examination may help determine the type, color, and composition of various fabrics indicating the nature of garment or fabric from which they originated.

Fibers are best handled with tweezers or smooth-jawed forceps. Place each fiber into a separate glass tube, vial, or bindle noting the location where it was found in addition to the patient name, date of birth, date, time, and initials of the person retrieving the fiber. For large threads or fibers, pick up with gloved fingers or forceps and place in a paper bindle. Place the vial or bindle in a clean outer envelope. Regular letter envelopes work well for this purpose as long as the edges are sealed. Seal and label the outer envelope appropriately. Affix a completed chain of custody tag.

Never place loose fibers directly into an envelope, since they can be lost. Bindle or otherwise contain the fibers first. Make a small bindle out of clean white paper if a glass vial is not available. A clean coin envelope may also serve as the inside container. Label the envelope before putting fiber inside, fold ends over, and tape closed.

Always package and send clothing of the patient from which fibers or threads might have originated to the forensic laboratory for comparison purposes. Individually package each item of clothing on which fibers and threads were found. Make a note on the outside of the envelope as to where fibers were seen.

If a suspect has been apprehended, advise law enforcement that you have collected representative fibers from the victim and suggest that they may want to do the same from the suspect's clothes, bedding, residence, or vehicle, and submit to the crime laboratory for comparison.

Hair

Hair should be collected by picking up with tweezers, placing in a labeled bindle or capped glass vial, and then into a labeled, sealed envelope with a chain of custody label attached to the outside.

When collecting hair samples from a victim, guidelines similar to those of the United States Drug Testing Laboratories, Inc., are suggested. Those instructions are found in Appendix 3.3.

Paint

Paint fragments or smudges may be found on skin, under fingernails, in hair, and on clothing. Paint fragments may be visible to the naked eye or they may be so small that they can only be detected by microscopic examination. Examine all areas, paying particular attention to those showing pressure, glaze, tears, or other signs of contact. Be alert to the possibility of paint transfer to clothing in pedestrian–motor vehicle accidents because its presence is usually minute.

Glass vials or cardboard pill boxes should be used to collect paint chips, since they protect paint from breakage and damage. If cardboard pill boxes are used, seal the box to prevent spillage of small paint samples. Never place paint samples into envelopes without bindling first.

- Small paint samples: Handle sample as little as possible. Carefully place samples in clean paper folded in a bindle. Place in a small clean envelope. Seal and package appropriately.
- Large paint chips: Handle sample as little as possible. Carefully place in a pill box or other clean protective container to protect from breakage. Seal and package appropriately.
- Paint on clothing: If paint is found on clothing, do not remove the paint. Carefully wrap each item separately by rolling it in paper, and placing each garment in clean separate paper bag. Seal and package appropriately. Make a note on the outside of the bag regarding visible or suspicious areas where paint might be found.

Metal or Sharp Items

To protect those handling evidence, knives, broken glass, or other sharp objects should be secured so they are unable to pierce the sides of the container. Small, sharp glass fragments, wire, and other such objects may be saved in a test tube or glass specimen jar. Avoid securing objects with tape. Paper or plastic-coated twist ties may be used to secure the item to a piece of

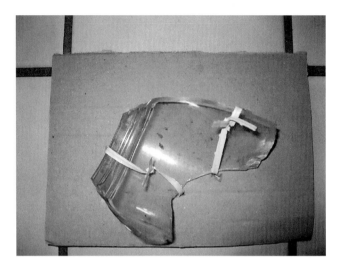

Figure 3.4 Glass packaging.

cardboard as follows. Puncture small holes in a piece of cardboard, place the twist tie through one hole around the object to be secured, back through a second hole in the cardboard, and twist together to secure (see Figure 3.4).

Glass

Fragments of glass may be found on personal belongings of suspects or victims involved in various types of crimes. Collect all glass present on clothing because more than one type may be present.

Place small glass fragments in paper bindles, then in clean coin envelopes, pill boxes, or film cans. Seal and label appropriately.

Place large glass fragments in clean, solid protective container boxes. Separate individual pieces with cotton or tissue to prevent breakage and damaged edges. Seal and label appropriately.

If glass is present on shoes or clothing of the suspect or victim/patients, remove shoes and clothing carefully, handle gently, fold, wrap in paper, and place in clean paper bag. Seal and label appropriately. Submit to the crime laboratory for examination.

Do not freeze metal or glass evidence items with blood or other body fluid stains. Rather, submit these items to law enforcement or the laboratory as soon as possible. As with all items of evidence, do not leave them unattended or unsecured at any time.

Pollen

Soil, dirt, dust, and hair contain abundant pollen and spores.

The goal in pollen evidence collection is to find a match between the pollen in a known geographical region with the pollen in a forensic sample, linking the victim or suspect with the location where the crime occurred. These samples must be collected carefully, gently, and without contamination. Always use sterile gloves and sterile collection tools when collecting pollen evidence to ensure that ambient pollen does not contaminate the sample.

When collecting soil pollen samples, advise law enforcement to collect control samples from the scene. Control samples are specimens of surface dirt from the region where a crime is believed to have been committed. The pollen recovered from the forensic specimen can then be compared to the crime scene pollen to see if they match.

Guidelines for Collecting Pollen Samples
- Make sure that all collecting tools and all collection containers are sterile.
- Avoid using contaminated tools.
- Place samples in sterile, paper containers, tightly sealed.
- Always collect more sample than is needed.
- If pollen is attached to clothing, do not remove from clothing. Place clothing in sterile paper bag. Label and seal appropriately.
- If pollen samples are soil, dust, or dirt, advise law enforcement to collect several control samples from the scene where the crime was committed.

Biological Evidence

Do not place items containing biological evidence in any type of plastic container! Bags containing wet items that may leak through the bag may be placed *on* a metal, glass, or plastic surface, but never *inside one.*

Body Fluid Collection

Control Swabs

The control swab provides the crime laboratory with a baseline of the patient's own secretions or possible contaminants adjacent to the stained area.

Collect a control swab by moistening a clean cotton applicator or swab with sterile or distilled water and swab an unstained area adjacent to the stain. For example, if the stain is on the right arm, collect the stain from an unaffected area near the stain on the same arm. Collect one control swab for each stain collected, unless several stains are collected within a small area. In that case one control swab is sufficient. Package each control swab separately. All evidence and control swabs must be air-dried prior to packaging.

Dried Blood

When dried, blood retains its usefulness as evidence indefinitely. Drying is, in fact, the preferred method of preserving body fluids. Blood type and DNA can be determined from dry samples. Dried blood should be collected with a clean cotton cloth or swab that has been moistened with sterile water. Leave a portion of the cloth or swab unstained as a control. Air-dry the cloth or swab and pack it into a clean paper bindle or an envelope with sealed corners. Do not use plastic containers.

Moist Secretions (e.g., nondried blood or other moist substances; semen or unknown liquid)

Absorb the material onto a clean, dry cotton cloth or cotton-tipped applicator swab to avoid dilution of the specimen. Gently roll the applicator or swab over the affected area(s); begin at the exterior of the area and move inward. For larger stains, collect the entire stain, using several swabs held together as a unit. Air-dry the specimen and pack it into a clean paper bindle or place it into an envelope or box with sealed corners. Do not use plastic containers. Label the specimen and seal with tamperproof tape.

Dried Secretions

For all dried specimens or secretions, slightly moisten a clean cotton applicator (for small areas) or gauze (for larger areas) with sterile distilled water. Gently move the swab inward over the injured or affected area(s) beginning at the outside edge. Label and air-dry prior to packaging. Place in an envelope, date, seal with writeable tape, and sign. It is important that all specimens be allowed to dry completely before packaging.

Urine Samples

Collect the first available urine sample. The forensic examiner always observes the collection process. If the patient must urinate prior to or during the evidence collection, collect the urine specimen at that time. Containers must be clean and tightly sealed in plastic or glass containers.

Have the patient void directly into the urine specimen bottle provided and fill the specimen container, if possible. After the specimen is collected, replace the cap and tighten to prevent leakage. Label with the patient's name, ID number, date of birth, date, time, and initials of person collecting the sample. Place the sample in a clean, leak-proof plastic bag, and attach chain of custody form. Note specimen collection in the nurse's notes of the patient's chart.

Gastric Contents

For collecting urine or gastric contents, use sterile containers for collection when feasible. If you are collecting these fluids in an uncontrolled scenario (e.g., the patient vomits onto the floor or is incontinent), use a clean towel to absorb the fluid. Ensure that one portion of the towel is maintained clean and dry as a control segment. After air-drying, package the towel(s) in a clean paper bag or large envelope. In some instances (e.g., items that are impractical to dry or if they are suspected of containing toxic substances), place them into a glass jar or paper bag placed on top of (not in) a plastic bag. Because plastic traps moisture and heat, damp materials are susceptible to mold, mildew, decay, or biological degradation.

Bite Marks

Human bite marks generally imply abuse, but not always. Bites from children are generally smaller than those from adults (see Figures 2.16 through 2.18). Animal bites are V-shaped lacerations and are almost always multiple.

Forensic odontologists (forensic dentists) can make an imprint of a bite mark as well as one of the teeth. Bite marks are uniquely identifiable (positive evidence), much the same as a fingerprint. Because the inflammatory process continues for some time after initial infliction of the bite, identification of a bite mark may be made several days after it is inflicted. An assailant can be identified by either a bite imprint on the victim, or by the victim's bite on an assailant. The most common location of bite marks are the back, breast, thigh, and buttocks.

An attempt should be made to recover secretory substances such as seminal fluid and saliva. Use moist swab techniques described in this chapter. These fluids are also observable under alternative light sources such as the Woods lamp.

Due to heavy bacterial loads in saliva, bite marks may lead to local and systemic infections. All bite marks must be thoroughly cleaned and aggressively managed by a medical team after evidence has been collected.

Blood Alcohol and Other Toxicological Specimens

Specimens collected in the course of medical treatment are not necessarily admissible for forensic purposes. Distinct forensic specimens may be required. There is currently debate on this subject, so it is advisable to know your local laws and practice. Your local prosecuting attorney can be contacted for current information. Standard policies and procedures should be in place and reviewed annually. In many states, a written request for obtaining the sample must be received from law enforcement, and a duly authorized forensic laboratory must analyze the specimen.

Common blood samples taken are:

- Alcohol and drugs—gray top vacuum tube (contains sodium fluoride)
- Serology
- DNA—yellow top vacuum tube
- Pregnancy, HIV

If an authorized forensic laboratory is not available, national labs can be found on the Internet.* Many specimens can be sent by certified mail. For instance, a forensic blood alcohol profile from United States Drug Testing Laboratory (USDTL) (BloodStat) identifies usage over a one- to two-day period and has the ability to provide under-the-influence interpretation. Their PEthStat will produce a positive result for an individual who has consumed one drink per day for 6 to 7 days a week. Their drug panels include amphetamines, cannabinoids, cocaine, opiates, phencyclidine (PCP), benzodiazepines, and barbiturates, plus methadone and propxyphene. It also includes the approximate number of drinks consumed as well as the blood alcohol concentration (BAC). A gray top tube containing at least 2 ml of blood should be submitted with a chain of custody form. USDTL suggests collecting the sample using an alcohol-free skin preparation.

In addition to drug testing services, USDTL provides addiction program support, research partnerships, child and family services, and transplant patient monitoring and a list of CPT codes for their procedures.

NMS Labs has over 2,500 standard blood, urine, and other fluid tests, including blood alcohol screening and testing,† testing for alcoholic content in beverages, and testing for flunitrazepam (Rohypnol—a date-rape drug). They even do blood spatter-pattern interpretation. Their Web site address is: http://www.nmslab.com.

---

* The following comments are for informational purposes only, and although they are all reputable laboratories, they are not necessarily a recommendation from the author.
† A forensic blood ethyl alcohol profile includes the approximate number of drinks consumed as well as the blood alcohol concentration (BAC).

Drug Detection Laboratories (DDL) can retest urine, blood, and drug samples for criminal prosecution, or may be used to check quantitative accuracy of samples previously tested by other laboratories. They can be found on the Internet at http://www.drugdetection.net. They offer forensic toxicologists who will qualify as experts in all aspects of the testing and interpretation of results.

Although most facilities have a distinct procedure for drawing specimens for blood alcohol levels, there are some basic principles and concepts that nurses need to understand. It is always a good idea to draw an extra purple top and red top tube in case they are needed at a later time. The first specimens obtained upon admission are the most likely to contain usable evidentiary data.

Note: Procedures as outlined in this section are relevant to any toxicology specimens drawn for forensic purposes in the healthcare setting. Samples taken for medical reasons are *not the same* as those taken for forensic reasons and are handled differently, including having a chain of custody.

Collect toxicology samples as soon as possible. Drugs may have been used during the commission of the crime and the presence of drugs or alcohol in the body will have legal significance. Many drugs are quickly eliminated from the body. Vials containing liquid blood samples should be refrigerated rather than frozen. Cleanse the blood collection site with a nonalcohol solution. Always fill the blood vial, if possible. Immediately after the blood collection, ensure proper mixing of the anticoagulant powder by slowly inverting the blood tube several times. Do not shake vigorously. Label appropriately.

An alcohol blood test must relate to an offender who committed a crime while under the influence of alcohol or while intoxicated.

If the patient is on anticoagulants or has a known blood abnormality such as hemophilia, he or she may be exempt from an ordered blood draw. When there is any question, obtain clarification prior to taking the specimen.

When obtaining the specimen, the usual procedures for a blood draw are followed with one exception. Use an alternate skin antiseptic such as providone iodine or benzalkonium to prep the skin prior to puncture. Note in the medical record that an alternate skin antiseptic was used. Although the use of an alcohol skin preparation does not adversely affect the value of the forensic sample, defense attorneys frequently dispute the findings on the basis that higher alcohol levels might be due to presence of medical alcohol on the skin and its inadvertent entry into the sample. Ethyl alcohol, however, can be distinguished from other forms during laboratory analyses. It is always prudent to eliminate any unnecessary courtroom disputes in the admissibility or credibility of any evidence. Therefore, it is advisable to specifically note that no alcohol was used in skin preparation.

If a patient refuses to have a venipuncture for blood alcohol level determination, the procedure should not continue until a subpoena or court order has been issued. Restraint and force are never appropriate to obtain a forensic specimen in a healthcare setting, even in police presence.

Law enforcement officers are both witnesses to the procedure and custodians of blood alcohol specimens. When the sample is obtained, (usually 10 ml) it should be sealed with tape and labeled with the name, date, time drawn, and the initials of who performed the phlebotomy. It should be given directly to the witnessing officer for transport to the state-approved laboratory. This information should also be precisely recorded in the patient's medical record.

In serious cases such as negligent homicide, two samples may be requested. Taken one hour apart, they help determine the alcohol level at the time of the incident or crime. Either the blood alcohol level will have peaked and started down, or it will be higher, indicating that the subject consumed alcohol immediately prior to the incident. It is estimated that the blood alcohol level falls 15–20 mg/100 ml per hour. For instance, if a blood alcohol sample is drawn four hours after an arrest and tests at 0.18% (180 mg/100 ml), it can be assumed that the subject's blood alcohol level at the time of arrest was approximately 0.26% or 260 mg/100 ml.

Testing for Drugs of Abuse

Medicines, controlled substances, marijuana, and drugs of abuse found in the possession of patients involved in various crimes can be analyzed by the crime laboratory. In cases where prescriptions are involved, the label itself provides valuable information regarding what should be found in the container, who prescribed the medication, when, and for whom the prescription was written. Therefore, leave medicines and other drugs in the original containers. Do not remove any marking or labeling.

If loose drugs (e.g., crack cocaine) are found, bindle and place in a prelabeled paper envelope. Place in a clean paper bag or container. Label the bag or container and seal using standard procedures. Seal loose materials securely, as they may spill or seep out of the container.

National laboratories can do forensic analysis/exclusion screens for many substances and drugs of abuse with one specimen. These *Rapid Tox* screens can test for amphetamines, sedatives, hallucinogens, hypnotics, narcotic analgesics, stimulants, and other substances including barbiturates, benzodiazepines, cannabinoids, cocaine and its metabolites, methadone, methanol, and opiates, as well as distinguishing between ethyl alcohol, methyl alcohol, and isopropyl alcohol. Toxicology screens generally require 8–10 ml of blood in a gray top (Fluoride Oxalate) tube. These specimens are stable for 10 days at ambient temperature and up to 20 days if refrigerated. Similarly, urine can be used for Rapid Tox screens with as little as 4–6 ml of urine. Urine specimens are stable for up to 7 days if refrigerated (NMS Laboratories, http://www.nmslab.com/).

Poisoning

Myriad substances, including lead, carbon monoxide, and some plants, can cause illness and death. They may be ingested, inhaled, or injected into the human body. Medicines, cleaning solutions, and other toxic materials are commonly found in the home. Keep all such substances out of the reach of children. Read labels carefully. Make sure all containers are clearly labeled and do not mix medicines or chemicals. Safely dispose of unused, unneeded, or expired prescription drugs. Do not treat medications as "candy"!

Poisoning can be self-inflicted or meted out by someone else. It can be accidental or intentional.*

If the victim has collapsed, call 911 immediately, assess ABCs, start cardiopulmonary resuscitation (CPR), and transfer the patient to a life-support facility immediately.

If the patient is awake and alert, call the National Poison Control Center's phone number: 800-222-1222. They are open 24 hours a day, 7 days a week. Have the following information available:

- Victim's age and weight
- Container or bottle of poison, if available
- The estimated time of poison exposure
- Address where poisoning occurred

Careful inquiry by the forensic nurse may shed light on the cause of the poisoning. When the following symptoms are not clearly tied to a cause, poisoning should be considered.

- Abdominal pain
- Blue lips
- Burns around the mouth
- Chest pain
- Confusion
- Cough
- Diarrhea
- Difficulty breathing
- Dizziness
- Double vision
- Drowsiness

* According to the CDC (Poisoning in the United States, Fact Sheet), in 2005, 32,691 poisonings were intentional—5,744 were suicides, 89 were homicides. In 2005, poison control centers received about 2 million poisoning calls and in 2006 over 700,000 individuals visited emergency departments for unintentional poisoning.

- Fever
- Headache
- Heart palpitations
- Irritability
- Loss of appetite
- Loss of bladder control
- Muscle twitching
- Nausea and vomiting
- Numbness or tingling
- Seizures or convulsions
- Shortness of breath
- Skin rash or burns
- Stupor
- Unconsciousness
- Unusual or chemical breath odor
- Weakness

If the person vomits, clear the airway. Always wear gloves, and if possible, wrap a clean cloth or paper towel around your fingers before cleaning out the mouth and throat. Roll the patient onto his or her left side and keep the person there to help prevent aspiration. Save any emesis or plant parts so they can be inspected visually or tested by a laboratory.

Keep the patient comfortable and warm. If the poison has spilled on the person's clothes, remove clothing, brush off any visible powdered chemical that might become corrosive when combined with water, and then flush the skin with water.

Things Not to Do

- Do not give an unconscious victim anything by mouth.
- Do not induce vomiting unless directed to do so by a poison control center.
- Do not try to neutralize the poison with lemon juice, vinegar, or any other substance unless directed by the poison control center.
- Do not use a universal antidote.
- Do not wait for symptoms to develop if you suspect poisoning.

Most poisons can be tested for by national labs. Numerous Web sites can be found on the Internet that can test for obvious and clandestine poisons. These Web sites also provide a wealth of information regarding taking, packaging, storage, and shipping of samples. An example from U.S. Drug Testing Laboratories can be found in Appendix 3.3.

Information on how to prevent poisoning should be available in all healthcare settings. The CDC has free materials such as buttons and badges. Fact sheets on poisoning prevention and can be found on the Internet at http://www.cdc.gov/HomeandRecreationalSafety/Poisoning/index.html.

Fetal Drug Exposure

Fetal exposure to drugs has been linked to long-term health problems in newborns. There is now a fast and reliable meconium drug testing method for the detection of fetal drug exposure providing the opportunity for appropriate treatment and intervention.

Meconium sample testing is noninvasive and eliminates the need to collect a blood or urine specimen from a newborn. The analysis provides up to a 20-week gestational detection window and results can be obtained in 48 hours or less, allowing for rapid healthcare treatment and planning for both mother and child. Meconium drug testing not only provides evidence of fetal drug exposure, but can also provide documentation of fetal alcohol exposure. Meconium testing has rapidly become the gold standard for diagnosing fetal alcohol and drug exposure. It is available as a 5-, 7-, 9-, and 12-drug panel. Alcohol and oxycodone levels are also available as add-ons (United States Drug Testing Laboratories [USDTL], 2010; Wikipedia, 2010).

DNA—CODIS

The Combined DNA Index System (CODIS) is an electronic database of DNA profiles administered through the Federal Bureau of Investigation (FBI). The system lets federal, state, and local crime labs share and compare DNA profiles (Zedlewski and Murphy, 2006).

CODIS uses two indexes: the Convicted Offender Index, which contains profiles of convicted offenders, and the Forensic Index, which contains DNA profiles from crime scene evidence.

In the past several years, CODIS has added several other indexes, including an Arrestee Index, a Missing or Unidentified Persons Index, and a Missing Persons Reference Index.

A record in the CODIS database, known as a CODIS profile, consists of a specimen identifier, an identifier for the laboratory responsible for the profile, and the results of the DNA analysis (known as the DNA profile). Other than the DNA profile, CODIS does not contain any personal identity information—the system does not store names, dates of birth, or social security numbers.

CODIS has a matching algorithm that searches the various indexes against one another according to strict rules that protect personal privacy.

For solving rapes and homicides, CODIS searches the Forensic Index against itself and against the Offender Index. A Forensic-to-Forensic match provides an investigative lead that connects two or more previously unlinked cases. A Forensic-to-Offender match actually provides a suspect for an otherwise unsolved case.

The real strength of CODIS lies in solving no-suspect cases. If DNA evidence entered into CODIS matches someone in the Offender Index, a warrant can be obtained authorizing the collection of a sample from that offender to test for a match. If the profile match is in the Forensic Index, the system allows investigators—even in different jurisdictions—to exchange information about their respective cases. For this reason, collecting blood samples for DNA analysis can be critical.

Sexual Assault

In cases of sexual violence, anal evidence is collected before conducting the vaginal examination and evidence collection. Contamination of the anal site and destruction of other evidence is then avoided. If anal penetration is reported, always observe for rectal trauma. Take photographs of the area.

A standard protocol has been developed by the Office of Violence against Women and is available free of charge from them. It is also downloadable from the Internet from http://www.SAFEta.org or through the International Association of Forensic Nurses at http://www.iafn.org.

Specimens to consider collecting include oral, anal, and rectal swabs, vaginal and cervical samples, and condoms for trace evidence.*

After the Evidence Collection

After the evidence collection, make sure all items of evidence have been labeled, dried, packaged in the appropriate containers, and sealed.

Never, for an instant, leave the evidence unattended before placing it in a locked area.

Advise law enforcement that the packaged evidence is ready for transfer. Verify each piece of evidence with the law enforcement person receiving custody and note that the evidence has been verified and the individual's name and badge number in the nursing record.

---

* Condom trace evidence is valuable for many reasons. In cases of sexual violence, recovered trace evidence from condoms may provide evidence of penetration or may correspond to those used by a certain manufacturer helping to link a serial rapist to a certain brand of condom.

Provide the patient a change of clothing. Arrange for follow-up care for treatment of medical and psychological issues as indicated. Give patients information on support groups so they can get help after leaving the hospital. Arrange transportation for patients if needed.

Summary

Nurses' observations of patients can yield an abundance of information in addition to appearance and anatomy. The watchful eye, discerning mind, and active forensic antenna will help to prevent many a forensic patient from falling through the cracks. It is worth repeating that the chain of custody is equally important to the admissibility of evidence in a court of law. Without a solid, unbroken chain of custody, evidence and its analysis are worthless. Understanding the chain of custody and meticulous adherence to evidence collection procedures are powerful tools nurses can use to aid in the administration of equity and justice.

Appendix 3.1: Evidence Tag and Chain of Custody

Patient's Name: _____

Date of Birth: _____   Patient sticker here

Date: _____

Case or File Number: _____

Name (printed) of individual sealing this container: _____

Signature of individual sealing this container: _____

Evidence contained: _____

_____

_____

Released by: _____ Received by: _____ Date: _____ Time: _____

Released by: _____ Received by: _____ Date: _____ Time: _____

Released by: _____ Received by: _____ Date: _____ Time: _____

Released by: _____ Received by: _____ Date: _____ Time: _____

Released by: _____ Received by: _____ Date: _____ Time: _____

Released by: _____ Received by: _____ Date: _____ Time: _____

This container is item _____ of _____ .

Deliver to Forensic Lab as soon as possible! Completing this data is critical to the prosecutorial value of the enclosed evidence.

Appendix 3.2: Classic Bindle Fold Diagram and Instructions

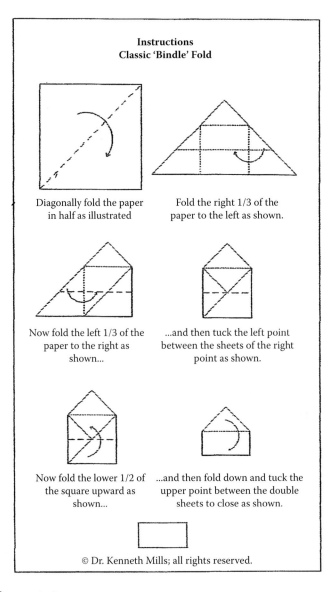

Instructions
Classic 'Bindle' Fold

Diagonally fold the paper in half as illustrated

Fold the right 1/3 of the paper to the left as shown.

Now fold the left 1/3 of the paper to the right as shown...

...and then tuck the left point between the sheets of the right point as shown.

Now fold the lower 1/2 of the square upward as shown...

...and then fold down and tuck the upper point between the double sheets to close as shown.

Used with permission.

How to Make a Bindle

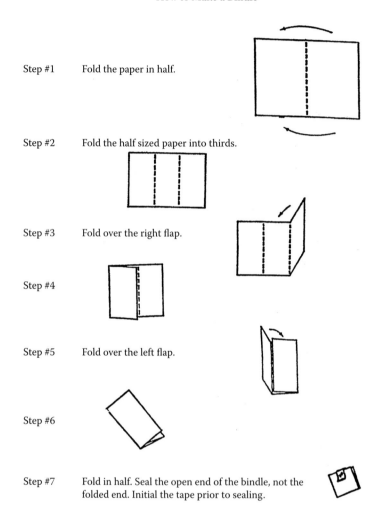

Step #1   Fold the paper in half.

Step #2   Fold the half sized paper into thirds.

Step #3   Fold over the right flap.

Step #4

Step #5   Fold over the left flap.

Step #6

Step #7   Fold in half. Seal the open end of the bindle, not the
               folded end. Initial the tape prior to sealing.

Appendix 3.3a: Blood Collection Instructions

Blood collection instructions

Materials needed for collection

- ▶ requisition form
- ▶ forensic blood collection kit
- ▶ courier exempt human specimen over-wrap

1. Verify the donor with a government-issued photo ID. (driver's license, state ID, passport)

2. Record the donor information on the requisition form.

3. Using one of the provided gray top Vacutainer tubes, execute blood draw following local Standard Operating Procedure. Discard the second Vacutainer tube if not needed.

4. Peel the long chain-of-custody label from the requisition form and affix over the cap of the transport tube. Have the donor initial and date the seal. **Failure to place label over the cap will result in a "Rejected Specimen".**

5. Have the donor print, sign and date the donor consent certification on the requisition form.

6. The collector should print, sign and date the collector certification on the requisition form.

7. Place the specimen tube(s) into the plastic tube holder.

8. Remove the adsorbent paper from the specimen bag and drape it over the tube between the two halves of the plastic tube holder.

9. Place the plastic tube holder in the specimen bag and seal the bag.

10. Place the requisition form and specimen bag into the exempt human specimen-labeled transport box and seal the box with the box seal sticker.

11. Place the transport box into the courier's exempt human specimen overwrap shipping bag. Contact your courier for pick-up. You may use the pre-printed UPS Label for overnight delivery to USDTL, call UPS for pick-up at (800) PICKUPS. A shipping fee will be invoiced per package.

USDTL
advancing the gold standard

1700 South Mount Prospect Road | Des Plaines, IL 60018 | (800) 235-2367 | www.usdtl.com

Used with permission.

Appendix 3.3b: Nail Collection Instructions

NailStat^{SM} collection instructions

Materials needed for collection

Provided by USDTL
- ▶ Envelope with security seal and foil inside
- ▶ Plastic specimen bag
- ▶ Requisition form / chain-of-custody (test subject ID number, date of collection, date and signature of sample collector must be completed at the time of collection)

Responsibility of the collector/collecting institution
- ▶ Metal nail clipper
- ▶ Alcohol and/or alcohol wipes (used to wipe clipper clean between collections)

1. Using a **clean** metal nail clipper, sequentially clip each fingernail collecting the nail clippings on the provided foil. Collect until 100 milligrams of nail clippings have been harvested.

2. Fold and cinch foil and place inside the envelope.

3. Seal the envelope with red security seal and initial seal. Place chain-of-custody number sticker from the requisition form on the envelope. The donor (guardian/witness/etc.) reads and initials the appropriate area on the envelope, then dates, prints and signs their name on the requisition form. The collector does the same. This initiates chain-of-custody.

4. In the presence of donor, place the white copy of the requisition form in outer pocket of security bag. Place envelope in other pocket of security bag. Then, seal and initial bag. The other copies of the requisition form are to be distributed at the discretion of the collecting institution.

5. Be sure that you indicate which NailStat^{SM} test(s) you require.

6. Contact your courier for pick-up. You may use the pre-printed UPS Label for overnight delivery to USDTL, call UPS for pick-up at (800) PICKUPS (742-5877). A shipping fee will be invoiced per package.

1700 South Mount Prospect Road | Des Plaines, IL 60018 | (800) 235-2367 | www.usdtl.com

Used with permission.

Appendix 3.3c: Cord Tissue Collection Instructions

CordStat™ collection instructions

1. Place name and I.D. number on the United States Drug Testing Laboratories (USDTL) requisition form and umbilical cord container.

2. Cut a 6-inch to 8-inch segment of the umbilical cord and drain and discard any blood.

3. Rinse the cord segment exterior with normal saline or an equivalent rinsing solution. **Important: avoid any contact of ethanol liquid or vapor with the umbilical cord.**

4. Place the umbilical cord segment into the umbilical cord container and sign the requisition form.

5. Place the barcode sticker from the requisition form over the top of the matching umbilical cord container.

6. Place the completed requisition form with tests ordered in the large pouch of the plastic specimen bag provided.

7. Place the umbilical cord container in the small pouch of the specimen and seal the bag.

8. Place the specimen bag in the cardboard box, close the box, affix the box seal at the indicated place and initial the seal.

9. Contact your courier for pick-up or UPS at (800) PICKUPS (748-5877). **Umbilical cord specimens should be refrigerated in a secure area until they are ready to be shipped.**

10. For more information on umbilical cord collection, visit www.usdtl.com and watch our CordStat℠ collection tutorial.

USDTL
advancing the gold standard

1700 South Mount Prospect Road | Des Plaines, IL 60018 | (800) 235-2567 | www.usdtl.com

Used with permission.

References

Assid, P. A. 2005. Evidence collection and documentation: Are you prepared to be a medical detective. *Topics in Emergency Medicine.* 27(1), 1–25. New York: Lippincott Williams and Wilkins, Inc.

Joint Commission on Accreditation of Healthcare Organizations (JCAHO or "The Joint Commission"). 2009. Family Violence, PC 01.02.09. http://www. JointCommission.org. Individual JCAHO recommendations cannot be accessed online. JCAHO can be reached at 630-792-5900.

NMS Labs. http://www.nmslabs.com

Quinn, C. 2005. *The medical record as a forensic resource.* Sudbury, MA: Jones & Bartlett Publishers, 156, 215.

USDOJ. 2010. Downloaded 8/23/2010 from http://www.dna.gov/dna-databases/ codis

USDOJ. 2010. Downloaded 8/23/2010 from http://www.fbi.gov/hq/lab/html/codis1. html

USDOJ, OJP, NIJ (2006).USDOJ, Washington, DC. http://www.ojp.usdoj.gov/nij/ journals/253 (accessed August 23, 2010).

United States Drug Testing Laboratories, d.b.a. MedStat Laboratories. Meconium testing. http://www.usdtl.com/mecstatsm.html (accessed August 23, 2010).

United States Drug Testing Laboratories, d.b.a. MedStat Laboratories. http://www. usdtl.com/medstat.html (accessed August 23, 2010).

United States Drug Testing Laboratories, d.b.a. MedStat Laboratories. Blood collection instructions. http://www.usdtl.com/uploads/users/files/collectioninstructions_Peh_%20FD_091809.pdf (accessed August 23, 2010).

United States Drug Testing Laboratories. Blood Collection. http://www.usdtl.com/ collection_instructions_and_training.html (accessed August 23, 2010).

United States Drug Testing Laboratories. Cord Tissue Collection. http://www.usdtl. com/collection_instructions_CodStat_%20FD_071009.pdf (accessed August 23, 2010).

United States Drug Testing Laboratories. Nail Collection. http://www.usdtl.com/ collection_instructions_Nail_Stat_%20FD_091809.pdf (accessed August 23, 2010).

Wikipedia. en.wikipedia.org/wiki/Combined_DNA_Index_System (accessed August 23, 2010).

Zedlewski, E. and M. B. Murphy. 2006. DNA analysis for "minor" crimes: A major benefit for law enforcement. USDOJ, OJP, NIJ Journal, (253), 4. Washington, DC.

Additional Resources

http://www.cdc.gov/ncipc/factsheets/poisoning.htm (accessed August 23, 2010).

http://www.ojp.usdoj.gov/nij/journals/253

http://crime.about.com/od/current/a/scott04718.htm

http://www.dna.gov/dna-databases/codis (accessed August 31, 2010).

http://www.fbi.gov/q/lab/html/codis1/htm (accessed August 31, 2010).

http://www.usdtl.com/uploads/users/files/collectioninstructions_PEth_%20 FD_091809.pdf (accessed August 31, 2010).

Domestic Violence and Abuse

4

Domestic violence is a pattern of assaultive and coercive behaviors including physical, sexual, and psychological attacks that adults or adolescents use against their intimate partners (Warshaw and Ganley, 1998)* (see Figure 4.1). The range of behaviors used by an abusive partner is extensive, and runs the gamut from intimidation to death. All are designed to increase the perpetrator's sense of power by exerting control over the victim, be it physical, sexual, emotional, financial, or a combination.

Patients generally fall into one of four categories:

- The patient has no history, disclosure, or pattern of behavior that suggests abuse.
- There is a history of abuse, but there is no abuse in the current relationship.
- There is current or recent abuse without physical injury.
- Abuse is currently or recently occurring and includes physical injury and/or threats of suicide or homicide.

Physical Abuse

Examples of physical abuse include:

- Hitting
- Kicking
- Slapping
- Punching
- Pushing
- Restraining or pinning to the wall, floor, or an object
- Hitting or breaking objects
- Threatening or actually hurting other people and/or animals
- Prohibiting access to medications (including birth control)
- Prohibiting access to healthcare

---

* As suggested by Warshaw and Ganley (1998), this is a behavioral definition rather than a legal one. You should check local statutes for the legal definition in your jurisdiction.

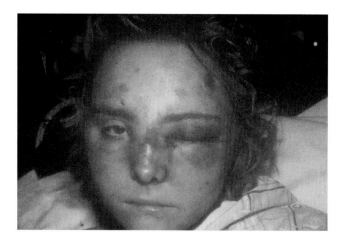

Figure 4.1 Domestic violence victim. Courtesy Dr. William Smock. Used with permission.

Sexual Abuse

Sexual abuse includes marital rape, including coercion into sexual activity under threat of violence or retaliation, forcing painful or uncomfortable sexual practices or positions, demanding the victim participate in unprotected sex, forced pregnancy or abortion, and forcing sexual conduct in front of children or other people.

Psychological Abuse

Psychological abuse includes the use of threats, intimidation, and ridicule as well as withdrawing attention and affection as punishment. Victims are often isolated from friends, family, church, work, and other social interaction. The perpetrator may also threaten to hurt or take away children, pets, parents, or other family members, and objects of personal value such as family heirlooms or special mementos.

Economic Abuse

Economic abuse includes limiting access to money by restricting employment, education, or other resources, incurring major debt, placing assets in only the perpetrator's name, and/or placing debt only in victim's name.

Most often, males are the perpetrators and females are the victims. Males are also victims of domestic violence, but the vast majority of victims are female (Tellez, Robinson, and Russell, 1999). The U.S. Department of Justice estimates that 95% of reported assaults on spouses or ex-spouses are committed by men against women (Douglas, 1991). Gay and lesbian partners are equally susceptible to being in an abusive relationship. The incidence is about the same as for heterosexual couples. For the sake of simplicity and ease of reading, victims will be referred to as female and perpetrators will be referred to as males.

An important aspect of domestic violence and abuse is the escalating nature and pattern of complete domination over time. A variety of acts and tactics are used to manipulate, intimidate, coerce, and terrorize the victim, exerting an ever-tightening net of control. When these tactics form a pattern, the relationship is no longer a "bad" or "dysfunctional" relationship, it is abusive. That is not to say that a single incident of violence is not abusive. It is, but violent behavior is rarely limited to a single event. Single events are merely a harbinger of the future. Any disclosure of interpersonal violence needs to be taken seriously.

Victims of domestic abuse are silent for many reasons. Often they are simply terrified. A significant aspect of abuse is that many tactics do not result in significant physical injury. Injuries resulting from physical assault are more easily recognized and can be debilitating or even life threatening. However, it is the psychological consequences of threats and coercion backed by the actual physical battering that result in profound and debilitating sequelae.

It is critical for the healthcare provider to understand that many victims of domestic abuse are unable or unwilling to leave the perpetrator immediately. We are obligated legally, ethically, and morally to make a good faith effort to identify victims of family violence and provide the victims with appropriate resources, giving them an opportunity to exit the cycle of violence when they are ready. We cannot, however, force the patient to make a particular decision or take a particular action. Much as we would like to rescue them, except for our legal obligations, the most effective thing healthcare providers can do is to show them kindness and understanding. Ultimately, an honest and trusting relationship between the healthcare provider and patient will serve the patient best.

Standards and laws to protect victims of domestic violence are now in place in most states. These vary from one jurisdiction to another, so it is advisable to review your local statutes, regulations, and ordinances. Failing to ask about violence in the home or living circumstances may expose the health practitioner to legal and regulatory liabilities. It is incumbent upon all healthcare provider institutions to develop policies and procedures to facilitate screening and intervention. Policies must specify the tools used for assessment, outline personnel responsible for tasks, list how the referral process is initiated and followed through, identify the elements of a safety plan, and describe how patient goals are to be met.

Joint Commission

The Joint Commission on Accreditation of Healthcare Organizations (JCAHO), or more recently referred to as the Joint Commission (JC), has created standards for screening and treating suspected victims of abuse. The healthcare provider must be able to demonstrate compliance with the intent of the standard (Scott and Matriccian, 1994).

Applicable Joint Commission standards are:

- Possible victims of abuse are identified using criteria developed by the hospital.
- Patients who are possible victims of alleged or suspected abuse or neglect have special needs relative to the assessment process.
- Administrators, business owners, government leaders, or healthcare providers ensure that the competence of staff is assessed, maintained, demonstrated, and improved continually.

The Joint Commission has noted that the intent of these recommendations is as important as the standards themselves. As noted by Scott and Matriccian (1994) it is incumbent upon the healthcare provider to demonstrate compliance with the intent of the standard. Criteria for meeting the standards should be developed by each institution. Criteria need to be objective and measurable. Individual patient differences, including general presentation, cognitive skills, needs (including language barriers),* disabilities, and emotional status need to be considered.

This means that staff must not only make an effort to identify victims of abuse, they must understand the extent and circumstances surrounding the current and past events and document it. Staff must not act in a manner that would encourage a patient to allege or fabricate abuse. Allegation alone is insufficient.

Information and evidence collected in the medical setting may have legal consequences. Special procedures to safeguard information and evidence must be in place, including how to initiate and maintain the chain of custody. Additional measures may be necessary for photography, special procedures, and release of information. Hospital policies must specify how information and evidence is to be handled and who is responsible. Documentation in the patient's record must include special consents

---

* Procedures should consider language barriers and the need for an interpreter. Interpreters should be available if the patient does not speak English. Family members do not make good interpreters, not only for domestic violence screening purposes, but for medical purposes because they may misunderstand or misinterpret legal terms used in the healthcare setting.

obtained, handling of evidentiary materials and chain of custody procedures, special legally required notifications, referrals made, lists and other
resource materials given to the patient, and information and/or evidence
given to law enforcement.

The Joint Commission recommends that staff be able to demonstrate competency in assessing, treating, and reporting victims of abuse. Simply remembering to ask is not enough. Records of staff education must be up to date and
demonstrate that education is ongoing and competence is routinely evaluated.

Intervention for Domestic Violence

- Inquire about abuse in a routine and standardized manner.
- Assess safety issues. See Appendix 4.1, "Danger Assessment—
  English" or Appendix 4.2 "Danger Assessment—Spanish."
- Treat medical and mental health issues.
- Discuss options and resources that are available.
- Provide advocacy and referral information.
- Document the abuse, referrals made, and treatments given.
- Provide for follow-up care (Warshaw and Ganley, 1998).

Patient privacy concerns must be addressed. Victims are reluctant to reveal
abuse because they are afraid that disclosure will result in repercussions
later. Perpetrators often accompany their victims to the healthcare setting
and are reluctant to leave the victim.* Before inquiring about abuse, it is
essential to create an environment where the victim feels she can talk freely.
Hospital policy may set up protocols so that the patient is initially seen alone
and accompanying individuals may join her later. Anyone accompanying
the patient (including another woman) should be considered a potential
perpetrator until ruled out. They may be charming and overly concerned
or defensive, controlling, abusive, or threatening. The healthcare provider
must create an environment of privacy that fosters disclosure and outlines
procedures to be followed if the immediate safety of the victim or staff is a
concern.

Healthcare settings should have posters and pamphlets visible and available free of charge. Post notices and support agency phone numbers regarding domestic violence on the inside of restroom doors or on the back doors of
ambulances and rescue units.

Perpetrators can be dangerous to healthcare staff as well as the victim.
Don't be intimidated by an overly protective partner. Think proactively. Have

---

* A red flag that domestic violence may be occurring is the demeanor and behavior of the
person accompanying the victim.

a plan in place to separate the patient from the perpetrator or other overly watchful companion. For your own protection as well as that of the patient, have a plan in place to alert security that a potentially volatile situation may be in the making. Staff should be able to lock doors quickly and easily, panic buttons should be available, alternate routes of escape should be established and clearly marked. Protocols for notifying security and or law enforcement should be in place and reviewed annually.

The Joint Commission recommends that staff be able to demonstrate competency in assessing, reporting, and treating victims of abuse (Mitchell, 2004; Scott and Matriccian, 1994).* Staff needs to become familiar with red flag conditions and suspicious circumstances. Simply remembering to ask is not enough. Policies and procedures must form the backbone of every patient's care.

A complete, up-to-date list of private and public resources should be developed and revised annually.

Acute medical injuries must be assessed in conjunction with overall health and safety needs. Immediate safety needs are paramount, both for the victim and for staff. Even in emergency situations, safety must be assessed simultaneously with medical needs.

Psychosomatic illnesses may occur. Depression, posttraumatic stress disorder, and suicide are common psychiatric implications. Drug and alcohol use has been shown to increase among victims once domestic violence begins. This increase is most likely a coping mechanism and a consequence of the abuse rather than the cause of the violence and abuse (Warshaw and Ganley, 1998).

Keep in mind that information and evidence collected during the screening, assessment, and treatment phases may be used later in legal proceedings. Policies and procedures must define the healthcare provider's responsibility for collecting, retaining, and safeguarding that information and evidence, including who is responsible for implementation and follow-through. In addition to assessment, treatment, and so on, documentation should include consents, chain of custody forms, legally mandated notifications, release of information to authorities, resource lists provided, referrals made, and plans for follow-up and other discharge planning activities. Check with the local and regional prosecuting attorneys for revisions to laws and practices.

---

* The three applicable standards are:
  1. Possible victims of abuse are identified using criteria developed by the hospital
  2. Patients who are possible victims of alleged abuse or neglect have special needs relative to the assessment process
  3. Leaders ensure that the competence of all staff is assessed, maintained, demonstrated and improved continually (Scott and Matriccian, 1994).

Screening

Basic screening is often done as part of initial intake and assessment procedures. The process is generally one of ruling out domestic violence. Those patients who provide negative answers to the initial screening can be exempted from further inquiry unless the interviewer has doubts about the truthfulness of the patient's responses.

Screening should be standardized and consistent so compliance and outcomes can be measured and evaluated. Chart prompts, check boxes, and standardized care plans are helpful and can save time.

Standardized screening questions should be straightforward and simple. Start with an introductory statement, such as the fact that about one-third of all women report having been a victim of domestic violence at some point in their lives. Tell the patient that routine screening is now a requirement for all health-care providers and that all patients are now asked these questions as a matter of routine. Inform the patient that their responses will be kept confidential.

Three basic questions suggested in the University of California, Davis guidelines are:*

- Have you been hit, kicked, punched, or otherwise hurt by someone within the past year?
- Do you feel safe in your current relationship?
- Is there a partner from a previous relationship who is making you feel unsafe now?

End with questions regarding violence in the home and patient concerns about violence or coerced/forced behavior (Mitchell, 1994).

The key is to look for a *pattern* of behavior both on the part of the perpetrator and on the part of the victim.

The following are considered to be clinical and patterned behavioral events that may be red flags for abuse:

- The history is inconsistent with clinical findings.
- The patient is young, isolated, or vulnerable.
- There is a history of stress-related illnesses or diseases.
- A history of mental health problems not otherwise explained and documented, including depression, a high level of anxiety, or suicide attempts.

---

* Basic screening does not ask about forced sex, emotional and psychological abuse, or about attempted strangulation. These issues can be explored if more in-depth questioning is warranted.

- There is a history of domestic violence and/or sexual assault in past as well as current living situations.
- There is suspicion of excessive or chronic alcohol or drug use, including prescription pain relievers or sedatives (especially when the diagnosis is absent or unclear or there are multiple healthcare providers).
- Physical injuries are central, bilateral, patterned, defensive, or inadequately explained.
- An injury occurs during pregnancy.
- The patient has avoided or been prevented from entering the healthcare system (including no treatment or inadequate treatment).
- The patient has a history of multiple minor or vague complaints, especially if the patient has been seen repeatedly.
- The patient defers healthcare decisions to the dominant partner.
- The patient reveals excessive or unexplained absences from work or other social activities.

If the patient denies abuse but her clinical presentation or behavior is highly suspicious for domestic violence, you may want to ask a second time. Use straightforward questions such as, "How did this happen?" or "Were your injuries caused by someone close to you?"

Be Prepared and Comfortable Dealing with a "Yes" Answer!

Key Point:
When assessing incidents of domestic violence, it is important to determine how long the abuse or threats have been occurring, and if there have been incidents of sexual or emotional abuse leading to mental health or physical consequences.

The presence of children in the home is particularly important. Any domestic violence safety plan must include any involvement of children—either as witnesses or as direct victims of abuse and/or neglect. Once children are identified, the next step is to evaluate the risk to children and seriousness of the violence on the children. Attempt to determine how they are being impacted and whether reporting is appropriate. Reporting of direct abuse of children is now required in all fifty states. Also note whether the patient has access to resources outside the home—work, social activities, access to family and friends, children's activities, and so on.

As healthcare providers, each of us needs to be aware of our own prejudices and biases that create barriers to an effective response.* Education and training that explore these issues should be a part of the standard protocol.

Victims of abuse may not report abusive situations due to one or more of the following factors:

- Concern that the batterer will retaliate with further abusive acts
- Social, cultural, and religious indoctrination that has instilled beliefs that marriage or other partnerships must be honored at all costs
- Age, disability, and language that reinforce that the individual is dependent upon the abusing partner

If your patient exhibits fear and begins to distance herself, reassure her that you are not a law enforcement officer and in most states are not required to report domestic violence. Offer the victim a card with local and national hotline numbers on it as well as basic information on developing a safety plan.

If screening is negative, accept it and move on to the standard intake and assessment. A Domestic Violence Assessment Tool in both English and Spanish is included in Appendices 4.1 and 4.2. Also included are a Domestic Violence Assessment Tool (Appendix 4.3) and an Interpersonal Violence Continuous Quality Improvement (CQI) tool (Appendix 4.4) for your facility. Created by Ms. Lisa Leiding for use in New Mexico, they can be adapted to fit other jurisdictions based on current law in your state.

If the answer is "no" and there is strong suspicions that domestic violence may be occurring, this information should be noted in the chart and the chart should be flagged for a situation reassessment at the next and every subsequent visit.† Explain to the patient that informational materials and resources are available at any time in the future, and explain what support is available locally, with precise information such as local shelters and toll-free hotline numbers.‡

Assessment

Once a domestic violence victim has been identified, a more in-depth evaluation needs to occur. Assessment requires an understanding of the dynamics

---

* Known barriers to effective response include the provider's own attitudes and misconceptions about domestic violence, apprehension regarding the length of time involved, familiarity with the patient, perpetrator, or their families, cultural assumptions and biases, belief that interpersonal relations are a private affair, and others.
† Chart prompts include notation that domestic violence (DV) screening was done: DV yes; DV no. Outcome of initial screening can be done using check boxes with DV+ (positive), DV– (negative), and DV? (suspected or questionable).
‡ The national domestic violence hotline number is: 800-799-SAFE.

of the relationship and the patient's level of understanding of the problem. In addition to physical and sexual findings, the clinician needs to assess contributing factors including personal history, family dynamics, support, and influence, as well as community values and support and other sociocultural values and influences.

Safety First

- Is the patient in immediate danger?*
- Is the staff in immediate danger?
- Where is the perpetrator?
- Where will the abuser be when the patient leaves?
- Does the victim want security or law enforcement to be notified?
- Are children and/or vulnerable adults safe?

Patient History

A complete history of recent events, particularly the one that precipitated contact with the healthcare system needs to be obtained. When, where, and how was the victim attacked? Have her describe the physical surroundings (bathroom, nearby woods, car, etc.).

Assess the pattern and history of abuse: How long has the abusive situation been going on? What techniques does he use to gain physical control? Has the intensity of abuse escalated either in frequency or severity?

Discuss the tactics used, including use of physical force, restraints, economic coercion, and psychological tactics such as threats of further injury and intimidation.

- Are drugs or alcohol involved?
- Can the victim predict violent episodes and has she developed any mitigating or preventive measures?
- Is there any sexual involvement?
- Are others also being abused or are they witnesses (children, vulnerable adults such as elderly or disabled parents, or others)?
- Does the abuser control the victim's activities, money, or access to the outside world?
- Does the victim believe she is being stalked? Has the abuser threatened to abduct her or actually taken her hostage for a period of time?

* A danger or lethality assessment should be done as part of the initial assessment.

- Was a weapon involved? What kind? Was the weapon threatened or actually used? How many times?
- Have there been threats of homicide or suicide?*

Identify the abuser by full name and his (or her) relationship to the victim. If known, document the abuser's current whereabouts. If there is current risk to the victim or healthcare provider, notify law enforcement. Inform the patient if law enforcement has been notified and why notification is necessary (i.e., legal requirement, patient and/or staff safety, etc.).

Assess the health impacts (physical, psychological, emotional, spiritual, etc.). Have previous incidents resulted in physical injuries? How is the abuse affecting the victim's overall mental and psychological health?

Assess the patient's understanding of the situation and willingness to leave. It is not uncommon for victims of domestic violence to be reluctant to leave their abuser. Despite orders of protection, the risk of serious physical harm or homicide *increases* once the victim leaves.

Determine appropriate resources and access to advocacy and support services. What personal resources does the victim have? Has the victim tried to use available resources in the past? What was the outcome? Is the victim willing or able to contact them again? What barriers does the victim have that might prevent using available resources (e.g., family, religious belief, cultural beliefs and/or practices, community or societal values)?

Prepare a *lethality* or *danger* assessment.† Include information and descriptions from the patient, if possible.‡

Determine whether children or other persons witnessed the event.

The patient's medical history is also important. It may reveal medical complaints without a specific diagnosis (headaches, heart palpitations, vague GI complaints, pelvic pain, chronic pain, insomnia, unexplained weight gain or loss, and other stress-related conditions).

Because of coercion and forced sex, the patient's reproductive history may include accounts of chronic pelvic pain, frequent vaginal and urinary tract infections, sexually transmitted diseases, pelvic inflammatory diseases, and HIV. Women victims of domestic violence are also prone to having an unplanned pregnancy, or a pregnancy with little or no prenatal care.

Pregnancy is also a time when women are particularly vulnerable to domestic violence. Domestic violence is a leading cause of maternal death (Fildes, Reed, Jones, Martin, and Barret, 1998).

---

* When a perpetrator says suicide, think homicide!
† See Appendix 4.1.
‡ Include information from other people sparingly, if at all. When used, clearly identify the person giving the information by name, relationship, and how they had possession of the information given. If the information is not firsthand, do not include it.

If the victim is pregnant, an obstetric consultation including a more thorough interview and exam is warranted. If blunt trauma to the abdomen is suspected, the workup should also include assessment of possible fetal, placental, or uterine injury. Was the pregnancy planned and welcomed by both partners? Was prenatal care postponed or totally absent? Have there been previous complications such as miscarriage, premature onset of labor, placenta abruptio, reports of decreased fetal movement, questionable rupture of membranes, or antenatal hospitalization?

Psychological History

Although acute physical injuries are the most obvious signs of domestic violence, it is the psychological sequelae that may be the most debilitating in the long term. There is a higher incidence of domestic violence among patients presenting with psychiatric issues. Interpersonal violence may exacerbate existing psychiatric illness or may actually be the cause of psychiatric difficulties. Long-term exposure to stress, depression, anxiety, and a feeling of hopelessness and helplessness may lead to tragic consequences.

A major goal of psychological assessment is to determine the patient's understanding and view of the problem. Does she (or he) understand that there really *is* a problem? Is she in denial? Try to determine the victim's general frame of mind. What, if any, communication and coping skills does the victim have when conflict arises? Is the victim rational or out of control? Where is she in the continuum between denial and readiness to leave?

Is the violence and attempt to control increasing? What effect has this had on the victim's mental and emotional status? What effect have other manipulative, threatening, or coercive acts such as attempted strangulation or presence of a gun in the house had on the victim? What is the nature and extent of threats made by the perpetrator? Is the perpetrator violent at work or in social situations? How does this impact the victim?

What is the level of intimidation or violence toward any children who are present? What impact does the violence have on the children?

Does the patient have psychosomatic complaints that cannot be explained? Examples are persistent headaches, chronic pain, GI problems, and musculoskeletal problems. Does she have stress-related symptoms, including increased anxiety, depression, posttraumatic stress disorder, changes in sleep or appetite patterns, fatigue or lack of energy, difficulty concentrating, sexual dysfunction, dizziness, palpitations, difficulty breathing, or unexplained paresthesias? Has she contemplated, threatened, or attempted suicide? Can she elaborate on her reasons?

Physical Examination

Ask the patient to remove all of her clothing as described in Chapter 3. To avoid further trauma to the patient, explain in detail the process of physical assessment and evidence collection, informing the patient about each step as you go.

Perform a complete, head-to-toe review of systems. Evaluate and describe each injury found, describing the type, number, size, and location. Mark each on a separate body map. Be particularly aware of injuries that support the narrative of the assault. For instance, injuries to the neck and throat might support an allegation of attempted strangulation (see Figure 4.2). Also note defensive wounds on hands, arms, and legs, and such details as hair, makeup, and torn clothing.

Acute injuries should be documented and treated. Included in a standard domestic violence assessment should be body diagrams and photographs. In addition to a standard head-to-toe physical examination, the forensic examination of a domestic violence victim includes assessment of the victim's demeanor, suicidal or homicidal ideation, and notation of the condition

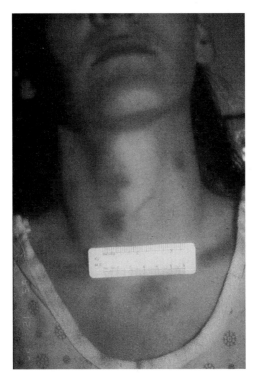

Figure 4.2 Fingertip contusions on neck. Courtesy Dr. William Smock. Used with permission.

and type of clothing worn and whether appropriate for weather and social environment.

Any injury to the face, eyes, mouth, or ears is suspicious for domestic violence. As with any repeated abuse, multiple injuries or bruises in various stages of healing are highly suspicious for domestic violence. Physical findings that warrant further inquiry about the possibility of partner abuse include injuries to the neck, face, breasts, abdomen, and any injuries to the genital or rectal area. Also note defensive injuries on the lateral surfaces of the victim's forearms or legs. As previously noted, injuries that are inadequately explained or are inconsistent with the injury patterns seen are highly suspicious for abuse. See Chapter 2, the section on patterned injuries* for things to look for and appropriate terms to use when documenting (Figures 2.8–2.20).

Victims may also be prohibited from seeking medical care for acute or chronic conditions. Chronic conditions such as diabetes, hypertension, heart disease, or other conditions such as epilepsy may be exacerbated in the presence of abuse. Victims of domestic violence may not take their medications as prescribed or be prohibited from doing so. Chronic pain is a common complaint among domestic violence victims. If the patient is taking medication, what is the diagnosis and who prescribed the medication? Are there multiple prescribing physicians?

Assess for conditions that could have been caused by prior acts of violence. These include hearing loss, headaches, back and neck pain, depression, or symptoms of posttraumatic stress disorder.

What is the patient's level of alcohol and drug use?

Intervention

Appropriate intervention for domestic violence includes the following:

- Inquire about abuse in a routine and standardized manner
- Assess safety issues
- Treat medical and mental health issues
- Discuss options and resources
- Provide advocacy and referral information
- Provide for follow-up care (Warshaw and Ganley, 1998)
- Document the abuse, referrals made, and treatments given

* Marks left by objects or burns from hot appliances or liquids.

Listen to the patient! Respond sincerely, thoughtfully, and nonjudgmentally. Let the victim know you understand. Validate her experiences and feelings. Acknowledge that the issues are complex and that it takes time to sort out emotions and make decisions. Support the victim's choices.

Advise the victim that she is not alone. Tell her you are there to help and resources are available when she is ready.

Inform the patient that domestic violence is a serious medical and psychological health issue. Advise her that acts of domestic violence are rarely static; more often than not they escalate in both frequency and severity over time.

Tell the victim she is not to blame. Domestic violence is not her fault, and responsibility for stopping it rests with the abuser.

Offer the victim referrals to local resources, and specifically address safe shelter, counseling, legal advocacy, and involvement of law enforcement. Provide the victim with a list of hotline numbers and local resources.

Work with the victim to prepare an individualized safety plan. Schedule follow-up per protocol. Document your findings and interventions.

Care Plan

Reassure the victim that she is not responsible and that help is available when she is ready. Provide appropriate information on resources that are available. Recognize that the victim may have feelings of shock, bewilderment, depression, fear, shame, denial, or isolation, and acknowledge that these are normal.

Immediate safety needs are a priority. Is the perpetrator there now or likely to be there in the near future? What response would she like from the healthcare staff? Does she want security or law enforcement to be notified? Does she have an order of protection? If so, does she want the abuser arrested? Where are children and other vulnerable family members? Does someone (family, friend, neighbor, etc.) need to be notified to pick them up? Will she go with the abuser if he wants her to leave? Is she prepared to go to a shelter or other safe place? Does she have the resources to do so?

Victims are often reluctant, unwilling, unable, or afraid to leave their abuser. Their safety and the safety of children and others in the household is a very real concern. Recognize, however, that it is unrealistic to expect the victim to immediately leave her situation. Family, friends, culture, finances, religious faith, and so on are barriers to a victim leaving a domestic violence situation. The average number of times a victim of interpersonal violence is seen before leaving the abuser is seven.

The care plan should address forensic (legal and social investigation) as well as medical, psychological, and safety needs. Included in the care plan should be narrative and photographic documentation and treatment for physical trauma or other unaddressed medical needs, assessment and care of mental health and safety needs, preparation of a safety plan and lethality checklist, and preparation of a follow-up plan that includes legal scenarios.

Inform the victim that battering is a crime and help from the legal system is available. Notify the victim early on whether reporting is mandatory in your area. Legal implications for the healthcare provider may involve a duty to report a battering assault to law enforcement and a duty to warn a third party if the provider is aware of a patient's, perpetrator's, or other individual's intent to harm them (Warshaw and Ganley, 1998, pp. 78–79). Battered women's shelters are usually aware of reporting requirements, police procedures, and consequences of reporting. Staff needs to be educated and knowledgeable in this regard so they can educate the victim at the time a safety plan is being developed. This is an area where a clinical forensic specialist is especially valuable.

Although reporting of domestic violence may or may not be mandatory, reporting of child abuse is. It is important to discuss the realities, benefits, hazards, and potential consequences of child protective enforcement services' involvement. It is *not* acceptable to return to an abusive situation if the children are also being abused or are in danger.

Appropriate intervention is dependent upon the care plan's specificity and thoroughness. Individuals responsible for each aspect of the care plan should be identified and listed. It is important to appreciate that the care plan is part of a comprehensive, collaborative, coordinated effort among healthcare, law enforcement, community services, victim advocacy organizations, and the legal community.

Care plans must include preparation for follow-up, including follow-up photographs, showing the progression of injuries over time.

Mental Health Plan

The patient's ability to cope and availability of a support system are critical to the success of the patient's psychological recovery. It is important that the practitioner understand the victim's ability to make decisions. A psychiatric referral may be necessary to adequately assess the patient's psychiatric future.

Severe depression, panic disorder, psychosis, and suicidal or homicidal ideation may require further psychiatric evaluation and treatment. Secondary psychiatric issues such as emotional, behavioral, and cognitive responses to abuse may be more a response to the abuse than true psychological illness. Once the victim is out of the abusive situation and safe from further violence, these issues may resolve.

Safety Plan

A detailed safety plan should be developed before the patient leaves. The patient's current circumstances, priorities, options, and readiness to leave will determine the specifics of the plan. Items to be addressed include the personal safety of the victim, other family members at risk, how to avoid and manage an explosive episode, how to identify and utilize resources, and how to prepare to leave if and when she is ready to do so.

Safety planning goals are based on whether any aspects of a safety plan are already in place. Find out what the patient has done to protect herself and her children. Determine whether she has made a realistic assessment of the extent and severity of the abuse. Assess what her current risk is for dangerous events and what she has done to protect herself and other family members.

Temporary housing is an immediate concern. List realistic options available include shelters, family, and/or friends. Or does she want to leave the community altogether? Does she consider this move temporary or permanent?

If the victim intends to return to the abuser, discuss what has worked in the past to reduce the level of violence. Does she believe previous strategies have been successful and has she thought of other strategies? Encourage her to inform her children to leave if possible and *not* to try to protect her. The best strategy is to arrange a safe place to meet the children later.

Does she have an escape (exit) plan for herself and her children? Has she discussed the exit strategy with them? What options are available to get her and the children out of the house and to safety?

Does she have a support network and how does she plan to use it in an emergency? Is there a friend or neighbor she can rely on if violence erupts or threatens to erupt? Would a family member or friend's presence in the home deter or mitigate a battering episode? Discuss other options she might have from her support network.

Is the patient able to anticipate a violent episode? If so, would she have time to leave before violence starts or escalates? If not, can she prepare a locked "delay" or "safe" room where she and children can buy some time while getting help or attempting to escape?

What precautions might she take to prevent escalation of violence? If the situation becomes violent, would she call the police? Could she call the police? Could children or others in the household call the police? Does she have an emergency signal established with a neighbor who could call for help? If not, could she develop one, and with whom?

Are there weapons in the home? How accessible are they? Can she remove them or remove ammunition?

Does she have a cell phone? (Landline phone lines can be cut.) Is it pre-programmed? To whom?

Encourage her to discuss the violence with her children, reassuring them that it is not their fault. Teach them how to use the telephone in an emergency and how to make collect calls should the abuser take them from the home.

Notify schools and those who provide child care with the names of those who may and may not pick up the children.

Discuss safety going to and from work and suggest she confide her situation with a sympathetic coworker, boss, or company security officer. Does she usually take the same route to work? Are there alternate routes or methods of transportation? Is she alone going to or from work? Can she arrange for someone to accompany her? Would security or someone else be willing to walk her to and from her car?

Encourage her to discuss the violence with a trusted friend or family member and discern who might be willing to help her in an emergency and to what extent are they willing to become involved.

Encourage her to read as much as she can about domestic violence. Remind her, however, that if such materials are discovered by the abuser, their presence might provoke an attack. Does she have or could she get a post office box and is there someone to whom she could give access to the mailbox?

Does she have a protective order in place? Would she be willing to get one? What would she do if the abuser violated a protective order?

Discuss safety measures such as changing locks on doors and windows, installing smoke detectors, having fire extinguishers readily available in several places throughout the house, and having outdoor lighting motion sensors installed.

What items would she need to have if she needed to flee suddenly? The following are items she might want to have sequestered away from the home.

- Medications and prescriptions
- Identification (driver's license or identification card, Social Security card, passport or green card, birth certificates for herself and her children)
- Important papers (marriage license, car title, lease/rental agreements, house deed or mortgage papers, insurance information and claim forms, school and health records, immigration papers, etc.)
- An extra set of keys
- Extra glasses
- Money, including change for a pay phone
- Checks or checkbook, banking books or account numbers, credit cards
- Protective orders, divorce and/or custody papers, and any other court documents
- Phone numbers of family and friends she can trust

- Phone numbers and addresses of shelters and other community or religious agencies
- Clothing and comfort items for herself and any children or dependent adults

If the victim is leaving the abuser, is there anyone with whom she can stay? Does she want to go to a battered woman's or homeless shelter? Are other temporary housing options available such as hotel vouchers?*

The most dangerous time for a victim of domestic violence is when she leaves. Does the victim want or need a protective order? What measures can she take to ensure it is enforced?

Does the victim want law enforcement to be notified?

If the abuser is leaving, suggest she take additional safety measures such as having locks changed, developing a secondary escape route (including second story egress such as a portable window ladder), installing motion sensors, security system, smoke detectors, and fire extinguishers.

Have friends, family, neighbors, coworkers, clergy, and others call 911 if they see the abuser near the house, work, school, church, or other place she habitually visits.

Resource List

All healthcare providers should maintain an up-to-date list of resources available in their community. That list should include the organization, its address, and the name and phone number of the contact person.

Community Resources

The following are resources within the community to which the victim can be referred:

- Emergency shelters
- Counseling and treatment programs
- Crisis providers
- Local and national hotlines
- Legal advocacy centers
- Education services
- Employment assistance

* Vouchers may be available from social services, advocacy programs, or your institution.

- Public awareness centers
- Child protective services
- Faith-based organizations and churches
- Alcohol and substance abuse counseling
- Support networks for the disabled and elderly
- Cultural organizations

Documentation

When documenting, remember to use direct patient quotes. Use the patient's own words to describe a discrete event, critical element, or history of abuse; *do not paraphrase victim statements.* Use the patient's own words to describe the health impacts of domestic violence. Write down the patient's utterances verbatim; do not change or alter words or word order. Remember that the patient's own words can be used in court as an exception to the hearsay rule and are therefore critical in the legal setting. Use quotation marks to enclose key elements or phrases to indicate that they are the patient's words.

Avoid use of the phrase "patient alleges"; use the phrase "patient states" instead. Avoid the use of judgmental evaluations or third-person or passive statements. Examples are "Patient overly excited" and "Patient struck in face." Instead, describe patient's emotional or mental status in observable terms. Use the patient's own words to describe who, what, when, where, and how.

Documentation specific to domestic violence needs to address all physical signs and symptoms, a danger assessment, and an assessment of risk for children as well as other adults in the household. Cultural, social, and familial considerations must be included. Note any special needs of disabled victims and actions taken on their behalf. All intervention or attempted intervention, referrals made or offered, and educational material provided should be clearly documented.

Documentation must include a safety assessment and plan, options discussed, whether police or other official agencies were notified, referrals made (including the name of the individual[s] taking the report), follow-up arrangements, and discharge instructions. List all of the information or literature regarding domestic violence that is given to the patient.

Appendix 4.1: Danger Assessment—English

DANGER ASSESSMENT
Jacquelyn C. Campbell, Ph.D., R.N.
Copyright 1985, 1988

Several risk factors have been associated with homicides (murders) of both batterers and battered women in research conducted after the murders have taken place. We cannot predict what will happen in your case, but we would like you to be aware of the danger of homicide in situations of severe battering and for you to see how many of the risk factors apply to your situation.

Using the calendar, please mark the approximate dates during the past year when you were beaten by your husband or partner. Write on that date how bad the incident was according to the following scale:

1. Slapping, pushing; no injuries and/or lasting pain
2. Punching, kicking; bruises, cuts, and/or continuing pain
3. "Beating up"; severe contusions, burns, broken bones
4. Threat to use weapon; head injury, internal injury, permanent injury
5. Use of weapon; wounds from weapon

(If any of the descriptions for the higher number apply, use the higher number.)

Mark **Yes** or **No** for each of the following. ("He" refers to your husband, partner, ex-husband, ex-partner, or whoever is currently physically hurting you.)

_____ 1. Has the physical violence increased in frequency over the past year?

_____ 2. Has the physical violence increased in severity over the past year and/or has a weapon or threat from a weapon ever been used?

_____ 3. Does he ever try to choke you?

_____ 4. Is there a gun in the house?

_____ 5. Has he ever forced you to have sex when you did not wish to do so?

_____ 6. Does he use drugs? By drugs, I mean "uppers" or amphetamines, speed, angel dust, cocaine, "crack," street drugs, or mixtures.

_____ 7. Does he threaten to kill you and/or do you believe he is capable of killing you?

_____ 8. Is he drunk every day or almost every day? (In terms of quantity of alcohol.)

_____ 9. Does he control most or all of the your daily activities? For instance: does he tell you who you can be friends with, how much money you can take with you shopping, or when you can take the car? (If he tries, but you do not let him, check here: _____)

_____ 10. Have you ever been beaten by him while you were pregnant? (If you have never been pregnant by him, check here: ___)

_____ 11. Is he violently and constantly jealous of you? (For instance, does he say "If I can't have you, no one can.")

_____ 12. Have you ever threatened or tried to commit suicide?

_____ 13. Has he ever threatened or tried to commit suicide?

_____ 14. Is he violent toward your children?

_____ 15. Is he violent outside of the home?

_____ Total "Yes" Answers

Thank you. Please talk to your nurse, advocate, or counselor about what the Danger Assessment means in terms of your situation.

From Jacquelyn Campbell, PhD, RN. 1985, 1988. Danger Assessment. Used with permission.

Appendix 4.2: Danger Assessment—Spanish

ESCALA DEVALORACION DE PELIGRO

En estudios realizados en mujeres que han muerto a consecuencia de violencia doméstica se han observado varios factores de riesgo tanto en el abusador como en la mujer golpeada. No podemos predecir que sucederá en su caso, pero me gustaría advertirle acerca del peligro de homicidio que se presenta en situaciones de agresión física severa con el fin de que usted se de cuenta de cuantos factores de riesgo se aplican a su situación.

(En las preguntas a continuación cuando hablamos de "el" nos estamos refiriendo a su esposo, compañero, ex-esposo o quienquiera que le esté haciendo daño físico en estos momentos)

Porfavor marque **SI** o **NO** a cada una de las siguientes preguntas.

SI NO

☐ ☐ 1. ¿ Durante el último año su pareja ha aumentado la frecuencia con que la golpea?

☐ ☐ 2. ¿ Durante el último año su pareja ha aumentado la severidad de la violencia física, ha utilizado un arma o ha amenazado con usarla?

☐ ☐ 3. ¿ Alguna vez ha tratado de estrangularla?

☐ ☐ 4. ¿ Hay alguna pistola en la casa?

☐ ☐ 5. ¿ Alguna vez su pareja la ha forzado a tener relaciones sexuales en contra de su volunted?

☐ ☐ 6. ¿ El utiliza algun tipo de drogas? por drogas me refiero a estimulantes, anfetaminas, "speed," polvo de angel, cocaína, crack, drogas que se venden en la calle, heroína, pastillas, inhalantes (thiner, cemento) cocktailes de drogas?

☐ ☐ 7. ¿ El la amenaza con matarla y/o usted cree que él es capaz de hacerlo?

☐ ☐ 8. ¿ El se emborracha diario o casi diario? (refiérase a la cantidad de alcohol)

☐ ☐ 9. ¿ El controla la mayoría de sus actividades diarias. Por ejemplo le dice quienes pueden ser sus amistades, cuánto dinero puede llevar cuando va de compras, cuando puede usar el coche, que no puede trabajar fuera de casa?

☐ ☐ 10. ¿ Alguna vez ha sido usted golpeada estando embarazada? (Si no era el padre de su hijo marque aqui _____)

☐ ☐ 11. ¿ El es extremadamente celoso al grado de portarse violento? (por ejemplo el le dice cosas como "Si no eres mía no vas a serlo de nadie")

☐ ☐ 12. ¿ Alguna vez usted ha amenazado o ha intentado suicidarse?

☐　☐　13. ¿ Alguna vez el ha amenazado o ha intentado suicidarse?
☐　☐　14. ¿ El es violento fuera de su casa?

Total de respuestas SI _____

GRACIAS POR SU COLABORACION. PORFAVOR HABLE CON SU MEDICO, ENFERMERA, ASESOR LEGAL O CONSEJERO ACERCA DE LOS RESULTADOS OBTENIDOS EN LA ESCALA DE VALORACION *DE PELIGRO* Y LO QUE ESTOS RESULTADOS SIGNIFICAN PARA SU SITUACION PERSONAL.

From Jacquelyn Campbell, PhD, RN. 1985, 1988. Danger Assessment. Used with permission.

Appendix 4.3: Domestic Violence Assessment Tool

INTIMATE OR DOMESTIC VIOLENCE
IMMEDIATE SAFETY

Where is your abuser now? _____
Do you feel that you are in danger now?
  □ No □ Yes

Where will the abuser be when we are done with the medical exam? □ Jail □ House □ Unknown
□ Other: _____
Hospital security notified? □ No □ Yes

VICTIM INFORMATION

BASIC INFORMATION

Height: _____
Weight: _____
Pregnant: □ No □ Yes □ Unknown
Ethnicity: _____

MEDICAL ASPECTS OF ABUSE

Are you allowed to have a primary care doctor?
  □No □ Yes
  Who: _____
Are you forced to seek care from a variety of providers or E.R.'s? □ No □ Yes
What prescription medications do you take? _____
Does your abuser hide or take your medications?
  □ No □ Yes
Do you drink? □ No □ Yes
Were you using during the violence? □ No □ Yes
Do you use illegal drugs? □ No □ Yes
Were you using during the violence? □ No □ Yes
Are you suicidal? □ No □ Yes
Have you thought about killing your abuser?
  □ No □Yes

HISTORY OF ABUSE

HISTORY

Who hurt you today? _____
Relationship: _____
How long have you known your
  abuser? _____
How long has the violence been going
  on? _____
Has the violence increased in frequency over
  the last year? □ No □ Yes
  How frequent? _____
Has the violence increased in severity over
  the last year? □ No □ Yes
Have others been harmed by your abuser?
  □ No □ Yes □ Unknown
If yes – what was their relationship to your
  abuser?_____

SEXUAL ASSAULT

Has your abuser forced or harmed you sexually?
  □ No □ Yes How long ago was the last
  incident? _____
  (If within 7 days stop the interview and call the SANE
  to examine).

STALKING

Have you ever felt trapped? □ No □ Yes
Has you abuser followed you? □ No □ Yes
Are you keeping written records? □ No □ Yes
Have you ever had an order of protection for your
  abuser? □ No □ Yes
Has anyone else ever filed an order against your abuser?
  □ No □ Yes □ Unknown
Has your abuser ever violated an order? □ No □ Yes

From Leiding, Lisa, RN, SANE. Interpersonal Violence C.Q.I. Tool for Medical Facilities. Used with permission.

ABUSE PATTERNS

PSYCHOLOGICAL ABUSE	ECONOMIC ABUSE	PHYSICAL ABUSE	PHYSICAL ABUSE	SEXUAL ABUSE
☐ Jokes	☐ Controlling money	☐ Harming pets	☐ Biting	☐ Sexual jokes
☐ Insults	☐ Stealing money	☐ Harming children	☐ Burning	☐ Use of pornography
☐ Namecalling	☐ Accusing victim of stealing money	☐ Denying basic needs	☐ Drugging	☐ Control contraception
☐ Crazy-making	☐ Misuse family income	☐ Destroying property	☐ Poisoning	☐ Withholding sex
☐ Jealousy	☐ Ruining victim credit	☐ Use of weapons	☐ Pinching	☐ Groping/ grabbing
☐ Isolation	☐ Forcing to work	☐ Using car as a weapon	☐ Poking	☐ Prostitution
☐ Abandonment	☐ Getting victim fired	☐ Driving recklessly	☐ Grabbing	☐ Forced sex with others
☐ Monitoring	☐ Not allowing to work	☐ Throwing objects	☐ Slapping	☐ Sex in front of kids
☐ Threats	☐ Making all decisions	☐ Being thrown	☐ Spanking	☐ Sadism
☐ Lying	☐ Destroying documents	☐ Restraining	☐ Punching	☐ Humiliation during sex
☐ Coercion	☐ Forcing criminal activity	☐ Strangulation	☐ Beating	☐ Beating after sex
☐ Humiliation	☐ _____	☐ Suffocation	☐ Whipping	☐ Beating during sex
☐ Sabotage-parenting		☐ Drowning	☐ Pushing	☐ Sex after beating
☐ _____		☐ Spont. Abortion	☐ Tripping	☐ Rape
		☐ Threat. Abortion	☐ Kicking	☐ Raped while drugged
			☐ _____	☐ STDs

SPIRITUAL ABUSE ☐ Insulting beliefs ☐ Making victim convert ☐ Using religion to justify abuse ☐ Not permitting religious practice

STAFF SIGNATURES

MD: _____

NURSE: _____   PATIENT'S STICKER

CURRENT HISTORY OF EVENTS IN PATIENT'S OWN WORDS

_____
_____
_____
_____
_____

SUBJECTIVE OBSERVATIONS DURING INTERVIEW PROCESS

_____
_____
_____
_____
_____

ABUSER INFORMATION AND FUTURE RISK

BASIC INFORMATION

Age: _____
Sex: ☐ M ☐ F
Pregnant: ☐ No ☐ Yes
Ethnicity: _____

MEDICAL HISTORY

Height: _____ Weight: _____
Medical history: ☐ None ☐ Unk _____
Mental history: ☐ None ☐ Unk _____
Does your abuser drink? ☐ No ☐ Yes
Does your abuser use drugs? ☐ No ☐ Yes
Does your abuser take prescription medications?
  ☐ No ☐ Yes
What? _____

LETHALITY

Has your abuser
  threatened suicide?
  ☐ No ☐ Yes
Has your abuser
  threatened to kill
  anyone? ☐ No ☐ Yes
Who? _____
When? _____
Have you asked for
  separation or divorce
  from your abuser?
  ☐ No ☐ Yes
  ☐ Not Applicable

WEAPONS

Does your abuser have
  access to a weapon?
  ☐ No ☐ Yes
  ☐ Unknown
Did your abuser use a
  weapon? ☐ No ☐ Yes

WORKPLACE VIOLENCE

Has your abuser come to your work
  place unexpectedly? ☐ No ☐ Yes
Does your abuser call your work
  place frequently? ☐ No ☐ Yes
Has your abuser ever hurt you in any
  way in a public place? ☐ No ☐ Yes

CHILDREN

Are there children involved?
  ☐ No ☐ Yes
Have children been hurt?
  ☐ No ☐ Yes
Have children witnessed the
  violence? ☐ No ☐ Yes

THREATS AND BEHAVIOR

Has your abuser lost a job or
  experienced problems at work?
  ☐ No ☐ Yes
Has your abuser displayed acts of
  extreme jealousy? ☐ No ☐ Yes
Has abuser threatened to take the
  kids away? ☐ No ☐ Yes
Has your abuser made threats to call
  immigration? ☐ No ☐ Yes
Has your abuser isolated you from
  family or friends? ☐ No ☐ Yes

PATIENT'S STICKER

Appendix 4.4: Domestic Violence CQI Tool

INTERPERSONAL VIOLENCE C.Q.I. TOOL
FOR MEDICAL FACILITIES
Created By: Lisa Leiding, RN, SANE

Hospital or Clinic:

GENERAL INFORMATION

Incident(s) CHECK
ALL THAT APPLY   Presentation

☐ Domestic Violence   ☐ January   ☐ February   ☐ Saturday   ☐ 0801 – 1600
☐ Intimate Partner   ☐ March   ☐ April   ☐ Sunday   ☐ 1601 – midnight
☐ Teen dating violence ☐ May   ☐ June   ☐ Monday   ☐ 0001 – 0800
☐ Sibling Abuse   ☐ July   ☐ August   ☐ Tuesday   ☐ Private Vehicle
☐ Parent Abuse   ☐ September ☐ October   ☐ Wednesday   ☐ Walk
☐ Child Abuse   ☐ November   ☐ December   ☐ Thursday   ☐ Ambulance
☐ Elder Abuse   ☐ Friday   ☐ Law Enforcement

REPORTING

MANDATORY REPORTING	AGENCY	WEAPONS	DISCLOSURE
☐ Child Abuse	☐ City Police	☐ Gun ☐ Knife	☐ Screened
☐ Elder Abuse	☐ County Sheriff	☐ Bat ☐ Car	☐ Self-disclosure
Interpersonal violence with:	☐ State Police	☐ Phone calls	☐ Suspected D. V.
	☐ Tribal Police	☐ Hands ☐ Feet	
☐ Shooting	☐ BIA	☐ _____	**PATIENT TYPE**
☐ Stabbing	☐ Children, Youth and	☐ _____	☐ Trauma (using
☐ Blunt Force Object	Families (CYFD)	☐ _____	trauma
☐ Unconscious Patient	☐ SANE		documentation)
☐ Death			☐ Abuser ☐ Victim

VICTIM INFORMATION

Ethnicity: _____   Injury Assessment: ☐ Physical Abuse
Immigration Status: ☐ Psychological Abuse   ○ Bruises ○ Scratches ○ Abrasions
☐ Documented   ☐ Economic Abuse   ○ Lacerations ○ Broken Bones ○ GSW
☐ Undocumented   ☐ Spiritual Abuse   ○ Burns ○ Bites ○ Internal Organ Injury
Age: _____   ☐ Sexual Assault/Abuse   ○ Head Trauma ○ Unconscious
                                                         ○ Abortion: Spontaneous or Threatened

Sex: ☐ Male ☐ Female   Victim Prescription Drug Use:
Pregnant: ☐ Yes ☐ No   ☐ None
☐ Not Known   ☐ _____

ABUSER INFORMATION   ### Substance Use

Ethnicity: _____   ☐ Same sex relationship   What was in system at time of incident?
Immigration Status: ☐ Relationship to victim   Victim Abuser
☐ Documented   _____   None   ☐   ☐
☐ Undocumented   ☐ Recent Separation/   Illegal Drugs   ☐   ☐
Age: _____   divorce   Alcohol   ☐   ☐
☐ Unknown age   ☐ Never Married   Abuser Prescription Drug Use:
Sex: ☐ Male ☐ Female   ☐ Children Present   ☐ None
Pregnant: ☐ Yes ☐ No   ☐ 1st Violent Incident   ☐_____
☐ Not Known   ☐ 2 + Violent Incidents   ☐_____
                             ☐ Violence in other
                             relationships

DISPOSITION AND REFERENCES

☐ Discharged   ☐ Transferred to: ☐ Restraining Order
☐ Home   _____   ☐ Legal Aide
☐ To friend's house   ☐ Shelter   ☐ Law Enforcement
☐ Admitted unit: _____   ☐ Follow-up Pictures
☐ Critical Care   ☐ Other: Counseling/Advocacy
☐ Med – Surg   _____   ☐ Rape Crisis Center
☐ Mental Health   ☐ D.V. Counseling

☐ Patient refused to discuss domestic violence. (Given hotline number and shelter information)

☐ Safety plan only

☐ Exam without photographs

☐ Full exam (examination, photographs, safety planning)

Identified problems in transportation, examination or   **FOR OFFICE USE ONLY**
  referral process: HealthCare Systems Implications:
_____   Private Insurance: ☐ Yes ☐ No
_____   Carrier _____
_____   Medicaid: ☐   Both ☐
_____   Medicare: ☐
_____   Self Pay: ☐ Yes ☐ No
                                                                Indigent Funds: ☐ Yes ☐ No

RECOMMENDED SOLUTIONS: **FOR OFFICE USE ONLY**
_____   How many prior visits? (1 year)
_____   Medical: _____
_____   Trauma related: ____
_____   How many subsequent visits?
_____   (1 year review)
                                                                Medical: _____
                                                                Trauma related: ___

Domestic Violence Team Comments or Suggestions: (EMS, MD, Nursing, Advocate, Shelter, Law
  Enforcement, Charge Nurse, etc.)

_____
_____
_____
_____
_____

PATIENT'S PRIMARY MD: _____ This patient has no Primary Doctor – made referral to: _____

STAFF	**PATIENT'S STICKER OR**
MD: _____	Patient Initials: _____ Date of Birth: _____
PRIMARY NURSE: _____	Date of Exam: _____ Account #: _____
CRISIS COUNSELOR: _____	Medical Record Number OR Employee
	Number: _____

This C.Q.I. tool must be removed from the Domestic Violence Flow Sheet after completed.
This tool will be routed to the Trauma Program Office ASAP. Statistics will be made available at
the completion of every quarter.

THIS FORM IS NOT PART OF THE PATIENT'S MEDICAL RECORD.

Confidential; Pursuant to Section 41-9-5, NMSA.

From Leiding, Lisa, RN, SANE. Interpersonal Violence C.Q.I. Tool for Medical Facilities. Used with
permission.

References

Douglas, H. 1991. Assessing violent couples. *Families in Society*. 72(9), 525–535.

Fildes, J., L. Reed, N. Jones, M. Martin, and J. Barret. (1998). Trauma: The leading cause of maternal death. *Journal of Trauma* 32(5): 643–645, referenced in Warshaw and Ganley. (1998). 56.

Joint Commission on Accreditation of Healthcare Organizations (JCAHO). (2009). Family Violence, PC 01.02.09. http://www.jointcommission.org (accessed August 23, 2010).

Mitchell, C. 2004. University of California, Davis Medical Center, California Medical Training Center, Connie Mitchell, M.D., Director, Domestic Violence Education. *Guidelines for the Health Care of Intimate Partner Violence for California Health Professionals*. Grant Award No. EM99041141 from the California Office of Emergency Services. Sacramento, CA.

Scott, C. J. and R. M. Matriccian. 1994. Joint Commission on Accreditation of Healthcare Organizations standards to improve care for victims of abuse. *Maryland Medical Journal*. 43(10): 891–898.

Tellez, T., K. Robinson, and M. Russell. 1999. Domestic Violence. *Topics in Emergency Medicine*. 21(2): 70–79.

Tjaden, P. and Thoennes, N. 1998. Prevalence, incidence and consequences of violence against women: Findings from the national violence against women survey. Research in Brief. Washington, DC: U. S. Department of Justice, National Institute of Justice. Also in 2002 National Victim Assistance Academy Text, Chapter 10. http://www.ojp.usdoj/ovc/assist/nvaa2002/chapter10.html (accessed August 31, 2010).

Warshaw, C. and A. Ganley, (eds.) 1995. *Improving the healthcare response to domestic violence: A resource manual for health care providers*. San Francisco: The Family Violence Prevention Fund.

Warshaw, C. and A. L. Ganley. 1998. *Improving the healthcare response to domestic violence: A resource manual for health care providers*. San Francisco: The Family Violence Prevention Fund, p.16.

Sexual Assault

5

Sexual assault (Kilpatrick, Edmunds, and Seymour, April 23, 1992) and abuse is any type of sexual activity to which the receiving individual does not agree. It can be verbal, visual, or anything that forces a person to participate in unwanted sexual contact. In addition to actual rape (vaginal, anal, or oral penetration), unwanted sexual contact includes inappropriate touching, child molestation, voyeurism, exhibitionism, incest, and sexual harassment (*Women's Health*, September 2010). Sexual assault is a crime.

Definitions

A clear, definitive definition of sexual assault is difficult to establish for a number of reasons. Social norms and expectations have changed over the generations. There are vast differences between localities and cultures. The United States federal statutes (U.S. Code, Title 18, 2010) are gender neutral, allowing for inclusion of male victims. They also do not limit criminal behavior to acts of penetration of the vagina by a penis, allowing for inclusion of other acts such as oral, anal, and digital penetration. Abusive sexual contact* includes behaviors such as intentional touching, including biting, of genitalia, anus, groin, breasts, inner thigh, or buttocks of any person (age and gender neutral) with an intent to abuse, humiliate, harass, degrade, or arouse, or to gratify the sexual desire of another. As stated in the *National Victim Assistance Academy (NVAA) manual* (2002), these federal definitions imply that the victim's state of mind at the time of the crime and the victim's physical and psychological injuries are important in helping to classify the crime and determine the appropriate punishment.

Goals of Care

The primary goals of patient care are to provide for immediate and long-term physical, psychological, emotional, and safety needs of the patient in a timely, sensitive, and supportive manner. Adjunct goals are more forensic in nature. They are to

---

* Note that the federal law does not use the term rape, nor does it require use of the term rape in order to meet the requirements of a crime.

- Encourage disclosure.
- Assist law enforcement in investigation by collecting high-quality, admissible evidence.
- Participate in a collaborative effort to determine the facts.
- Facilitate appropriate prosecution by cooperation with attorneys.
- Provide substantive, unbiased testimony at trial.

Although healthcare personnel see themselves as advocates for the patient in the medical setting, it is extremely important that they function as neutral, impartial collectors, protectors, and disseminators of the facts, and *nothing else*. Healthcare systems also have a legal and moral obligation to provide equivalent and unbiased healthcare services to the perpetrator.

Injuries

Serious physical injuries do occur and must be a priority in patient care. Most physical injuries, however, are minor—abrasions and contusions, minor cuts, and so on. The most consequential issue for the sexual assault victim is the devastating psychological impact. An intimate personal violation such as sexual assault can have a long-term emotional impact on the victim. For the victims of sexual assault, life will never be the same. Rape, or even attempted rape, leaves the victim feeling overwhelmed and vulnerable. Shock, anger, grief, and depression are common reactions. Victims are often confused and find it difficult to concentrate or make decisions (including the decision to cooperate with law enforcement or prosecution). Posttraumatic stress disorder (PTSD) is a serious and debilitating psychological disorder and can occur months or years after a traumatic event. Almost one-third of sexual assault victims develop PTSD at some point in their lives. In short, sexual assault is a psychological crisis rarely trumped by any physical trauma that may have occurred.

Essential elements of recovery include rapid psychological intervention and stabilization. It is vital for healthcare personnel to understand the important role of psychological issues in the long-term recovery of the sexual assault victim and to respond promptly by providing supportive care.

In addition, safety issues are a paramount concern. Sex offenders frequently repeat their offense. For a variety of reasons, many victims never report their rape to the authorities. Not only does the victim fail to receive justice (or achieve closure), the rapist remains unpunished and is able to victimize others (Kilpatrick, Edmunds, and Seymour, 1992). Most perpetrators of sexual assault are known to the victim and victims fear retaliation. Therefore, safety and confidentiality issues are extremely important to the victim. Additionally, according to the National Women's Study (Tjaden and Thoennes, November 1998), over two-thirds of sexual assault victims were under the age of 17—an age where emotional and psychological abilities are not yet fully matured and integrated.

Role of the First Responder and Healthcare Provider

First and foremost, determine that the scene, the patient, and you are safe. If the whereabouts of the assailant are unknown or likely to return, remove the patient to the rescue vehicle and leave the scene as soon as possible. Collect any bedding, clothing, and so on, according to chain of custody standards if possible, but do not delay getting yourself and the patient to a safe place.

In addition to standard emergency response measures, there are some things the first responder can and should do when first encountering the victim of sexual assault.* In addition to getting consent for treatment and advising the patient of his or her rights under Health Insurance Portability and Accountability Act (HIPAA), have the patient sign a consent to photograph form. Explain that the victim has the right to revoke permission at a later date, but immediate and follow-up photography are critical pieces of prosecution, even if prosecution is delayed.

If the scene is determined to be safe, take photographs as soon as possible using the basic techniques described in Chapter 2. Photograph the scene and victim exactly as found. Try to determine where the assault occurred. If it occurred in another room, photograph that scene as well. If the sexual assault occurred at another location, get as much information as possible as to the location and notify police so they can investigate. Collect and secure any evidence as explained in Chapter 3 or have law enforcement do it if they are available.

Wear appropriate personal protective equipment (PPE), especially gloves. Change them often to avoid cross-contamination. If you come into direct contact with biological fluids that may contain evidential material, you may air-dry the gloves and package as other evidence, labeling appropriately with the name of the responder and location biological material was encountered.

Encourage the patients to proceed with evidence collection, even if they are unsure whether they will cooperate with police or pursue prosecution. Evidence not collected is evidence that cannot be used if patients change their minds later. Because some evidence is time-sensitive and fragile, it is also important to document the time the assault allegedly occurred and the time the evidence was collected.

---

* Suggestions for police officers and first responders on how to interact with victims of sexual assault are found in Appendices 5.1 and 5.2.

Ask if there was any oral involvement during the assault. If so, do not give the patient anything to eat or drink until oral swabs have been collected.

Ask if the patient has showered. If so, using the evidence collection procedures listed above, collect all clothing the patient was wearing at the time, including shoes. If the patient has not showered, inform her that because there may be valuable evidence on her body, it is advisable she does not shower or change clothes. Collect any blanket or sheets that might have been under or around the patient. Have someone pack a bag with a complete set of clean clothes for the patient to wear home after the physical exam. Ordinarily, the patient will be offered the opportunity to shower as soon as the exam has been completed and evidence has been collected.

Give the patient a brief overview of what to expect: an initial interview (with or without police and rape crisis advocate present), a physical exam (including pelvic exam), screening for pregnancy and sexually transmitted diseases, and prophylactic treatment offered, photographs taken, and collection of evidence for prosecution (including clothes, hair, and body fluids). Inform patients that they can refuse any services offered at any time, and that they will not be responsible for the cost of the forensic exam.* Some states, but not all, will also pay for portions of the exam that are medical in nature (i.e., treatment for preexisting conditions or diseases).

Sexual Assault Nurse Examiners (SANE)

To meet the unique needs of victims of sexual assault, the sexual assault nurse examiner (SANE) or sexual assault forensic examiner (SAFE) programs were developed. *Examination by a SANE nurse is highly recommended.* National protocols for adults and adolescents have been written and are available free of charge as a download from the Internet or in CD form from the Department of Justice, Office of Violence Against Women (USDOJ, OVW, 2004).† Educational programs for SANE nurses have been created and certifications have been established. Because of the victim's special needs, the SANE nurse's higher level of education and expertise, and the collaborative relationship between law enforcement, advocacy, and the judicial system, it is *highly* recommended that all victims of sexual assault be referred to a SANE program as soon as possible.

Legal and prosecutorial issues are complicated. SANE nurses are trained in how to avoid common pitfalls in the process such as the issue of consent, jurisdictional issues, handling biological evidence, the difference between

---

* Make sure you know where billing should be sent in your jurisdiction.
† Note that these do not apply to prepubescent children as those exams are significantly different.

medical specimens and forensic specimens, and payment for treatment. It is strongly suggested that all healthcare providers read and understand the President's DNA Initiative National Protocol for Sexual Assault Medical Forensic Examinations.

A National Protocol

A National Protocol for Sexual Assault Medical Forensic Examinations of Adults and Adolescents (SAFEta) is available free of charge from the U. S. Department of Justice, Office of Violence Against Women (September 2004; NCJ 206554).

The following material is a replication of the Goals of the National Protocol and the Recommendations at a Glance (a summary) (USDOJ, OVW, SAFEta, 2004). The entire protocol is downloadable from the Internet or available free of charge in CD form. The author strongly urges the reader to acquire and review the information contained therein. (Note: footnotes have been styled for clarity and ease of reading. Page references have been left intact; they refer to specific pages in the National Protocol.)

Goals of the National Protocol for Sexual Assault
Medical Forensic Examinations

Consider what it might be like to be a victim of sexual assault who has come to a healthcare facility for a medical forensic examination. Sexual assault is a crime of violence against a person's body and will. Sex offenders use physical and/or psychological aggression to victimize, in the process often threatening a victim's sense of privacy, safety, and well-being. Sexual assault can result in physical trauma and significant mental anguish and suffering for victims. Victims may be reluctant, however, to report the assault to law enforcement and to seek medical attention for a variety of reasons. For example, victims may blame themselves for the sexual assault and feel embarrassed. They may fear their assailants or worry about whether they will be believed. A victim may also lack easy access to services. Those who have access to services may perceive the medical forensic examination as yet another violation because of its extensive and intrusive nature in the immediate aftermath of the assault. Rather than seek assistance, a sexual assault victim may simply want to go somewhere safe, clean up, and try to forget the assault ever happened.[*] It is our hope that this protocol will help jurisdictions to respond to sexual assault victims in the most competent, compassionate, and understanding manner possible.

This protocol was developed with the input of national, local, and tribal experts throughout the country, including law enforcement representatives,

[*] Paragraph adapted in part from the Ohio Protocol for Sexual Assault Forensic and Medical Examination, 2002, p. 2.

prosecutors, advocates, medical personnel, forensic scientists, and others. We hope that this protocol will be useful in helping jurisdictions develop a response that is sensitive to victims of sexual assault and that promotes offender accountability. Specifically, the protocol has the following goals:

- Supplement but not supercede the many excellent protocols that have been developed by States, tribes, and local jurisdictions, as well as those created at the national level. We hope that this protocol will be a useful tool for jurisdictions wishing to develop new protocols or revise their existing ones. **It is intended as a guideline for suggested practices rather than a list of requirements.** In many places, the protocol refers to "jurisdictional policies" because there may be multiple valid ways to handle a particular issue and which one is best should be determined by the jurisdiction after consideration of local laws, policies, practices, and needs.

- Provide guidance to jurisdictions on responding to adult and adolescent victims. Adolescents are distinguished in the protocol from prepubertal children who require a pediatric exam. **Pediatric exams are not addressed in this document. This protocol generally focuses on the examination of females who have experienced the onset of menarche and males who have reached puberty.** Legally, jurisdictions vary in the age at which they consider individuals to be minors, laws on child sexual abuse, mandatory reporting policies for sexual abuse and assault of minors, instances when minors can consent to treatment and evidence collection without parental/guardian involvement, and the scope of confidentiality that minors are afforded. **If the adolescent victim is a minor under the jurisdictional laws, the laws of the jurisdiction governing issues such as consent to the exam, mandatory reporting, and confidentiality should be followed.**

- Support the use of coordinated community responses to sexual violence. Although this document is directed primarily toward medical personnel and facilities, it also provides guidance to other key responders such as advocates and law enforcement representatives. This type of coordinated community response is supported by the Violence Against Women Act and subsequent legislation. Such a response can help afford victims access to comprehensive immediate care, minimize trauma victims may experience, and encourage them to utilize community resources. It can also facilitate the criminal investigation and prosecution, increasing the likelihood of holding offenders accountable and preventing further sexual assaults.

- Address the needs of victims while promoting the criminal justice system response. Stabilizing, treating, and engaging victims as essential partners in the criminal investigation are central aspects of the protocol. Thus, this protocol includes information about concepts such as "blind reporting," which may give victims needed time to decide if and when they are ready to engage in the criminal justice process. A blind report may also provide law enforcement agencies with potentially useful information about sex crime patterns in their jurisdictions. The objective is to promote better and more victim-centered evidence collection, in order to provide better assistance in court proceedings and hold more offenders accountable.

- Promote high-quality, sensitive, and supportive exams for all victims, regardless of jurisdiction and geographical location of service provision. The protocol offers recommendations to help standardize the quality of care for sexual assault victims throughout the country. It also promotes timely evidence collection which is accurately and methodically gathered, so that high-quality evidence is available in court.

This protocol discusses the roles of the following responders: healthcare providers, advocates, law enforcement representatives, forensic scientists, and prosecutors. Clearly, each of these professions has a role in responding to victims, investigating the crime, and/or holding offenders accountable. But rather than dictate who is responsible for every procedure within the exam process, the protocol is designed to help communities consider what each procedure involves and any related issues. With this information, each community can make decisions for its jurisdiction about the specific tasks of each responder during the exam process and the coordination needed among responders. The following is a general description of what each responder may assist with:

- **Advocates** may be involved in initial victim contact (via 24-hour hotline or face-to-face meetings); offer victims advocacy, support, crisis intervention, information, and referrals before, during, and after the exam process; and help ensure that victims have transportation to and from the exam site. They often provide followup services designed to aid victims in addressing related legal and nonlegal needs.
- **Law enforcement representatives** (e.g., 911 dispatchers, patrol officers, officers who process crime scene evidence, and investigators) respond to initial complaints, work to enhance victims' safety, arrange for victims' transportation to and from the exam site as needed, interview victims, coordinate collection and delivery of evidence to designated labs or law enforcement facilities, and investigate cases.
- **Healthcare providers** assess patients for acute medical needs and provide stabilization, treatment, and/or consultation. Ideally, sexual assault forensic examiners perform the medical forensic exam, gather information for the medical forensic history, and collect and document forensic evidence from patients. They offer information, treatment, and referrals for sexually transmitted infections (STIs) and other nonacute medical concerns; assess pregnancy risk and discuss treatment options with the patient, including reproductive health services; and testify in court if needed. They typically coordinate with advocates to ensure patients are offered crisis intervention, support, and advocacy during and after the exam process and encourage use of other victim services. They may follow up with patients for medical and forensic purposes. Other health

care personnel that may be involved include, but are not limited to, emergency medical technicians, staff at hospital emergency departments, gynecologists, surgeons, private physicians, and/or local, tribal, campus, or military health services personnel.

- **Forensic scientists** analyze forensic evidence and provide results of the analysis to investigators and/or prosecutors.
- **Prosecutors** determine if there is sufficient evidence for prosecution and, if so, prosecute the case. They should be available to consult with first responders as needed. A few jurisdictions involve prosecutors more actively, paging them after initial contact and having them respond to the exam site so that they can become familiar with the case and help guide the investigation.

This document is intended only to improve the criminal justice system's response to victims of sexual assault and the sexual assault forensic examination process and does not create a right or benefit, substantive or procedural, of any party.

Recommendations at a Glance: A National Protocol for
Sexual Assault Medical Forensic Examinations

The National Protocol for Sexual Assault Medical Forensic Examinations offers guidance to jurisdictions in creating and implementing their own protocols, as well as recommending specific procedures related to the exam process. *Recommendations at a Glance* highlights key points discussed in the protocol, but it is not designed to be a stand-alone checklist on exam procedures or responsibilities of each involved responder. The protocol should be read to understand and respond to the complex issues presented during the exam process. See the protocol introduction for an explanation of select terms used in this chapter and the protocol.

Goal of the Protocol

A timely, well-done medical forensic examination can potentially validate and address sexual assault patients* concerns, minimize the trauma they may experience, and promote their healing. At the same time, it can increase the likelihood that evidence collected will aid in criminal case investigation, resulting in perpetrators being held accountable and further sexual violence prevented.

The examination and the related responsibilities of health care personnel are the focus of this protocol. Recognizing that multidisciplinary coordination is vital to the success of the exam, the protocol also discusses the responses of other professionals, as they relate to the exam process.

---

* Sexual assault patients are also referred to as victims, depending on which responders are primarily being discussed. The term "patients" is generally used by health care professionals.

A. Overarching Issues

1. Coordinated approach: A coordinated, multidisciplinary approach to conducting the exam provides victims* with access to comprehensive immediate care, helps minimize trauma they may experience, and encourages their use of community resources. Such a response can also enhance public safety by facilitating investigation and prosecution, which increases the likelihood that offenders will be held accountable for their actions. Raising public awareness about the existence and benefits of a coordinated response to sexual assault may lead more victims to disclose the assault and seek help.

   Recommendations for jurisdictions to facilitate a coordinated approach to the exam process:
   - Understand the dual purposes of the exam process to address patients' needs and justice system needs. Addressing patients' needs may include evaluating and treating injuries; conducting prompt exams; providing support, crisis intervention, and advocacy; providing prophylaxis against sexually transmitted infections (STIs) and referrals; assessing reproductive health issues; and providing followup contact/care. Addressing justice system needs may include obtaining a history of the assault; documenting exam findings; properly collecting, handling, and preserving evidence; and (postexam) interpreting/analyzing findings, presenting findings, and providing factual and expert opinions.
   - Identify key responders and their roles.
   - Develop quality assurance measures to ensure effective immediate response.
2. Victim-centered care: Victim-centered care is paramount to the success of the exam process. Response to victims should be timely, appropriate, sensitive, and respectful.

   Recommendations for health care providers and other responders to facilitate victim-centered care:
   - Give sexual assault patients priority as emergency cases and respond in a timely manner. Provide them with as much privacy as possible, while ensuring that they are supported.
   - Recognize that the medical forensic exam is an interactive process that must be adapted to the needs and circumstances of each patient.
   - Be aware of issues commonly faced by victims from specific populations. For example, certain characteristics (e.g., culture, religion, language skills/mode of communication, disabilities, gender, and

---

* The term "victim" is not used in a strictly criminal justice context. The use of "victim" simply acknowledges that persons who disclose that they have been sexually assaulted should have access to certain services.

age) may influence a victim's behavior in the aftermath of an assault, including the exam process.

- Understand the importance of victim services within the exam process. Victim service providers/advocates typically offer victims support, crisis intervention, information and referrals, and advocacy to ensure that victims' interests are represented, their wishes respected, and their rights upheld. Providers/advocates also may offer support for family members and friends who are present. In addition, they can promote sensitive, appropriate, and coordinated interventions.

- Involve victim service providers/advocates in the exam process as soon after a victim discloses an assault as possible. Victims have the right to accept or decline victim services.

- Accommodate patients' requests to have relatives, friends, or other support persons (e.g., a religious/spiritual counselor) present during the exam, unless the presence of that person could be considered harmful. (See *C.4. The Medical Forensic History* for confidentiality considerations regarding the presence of these individuals during history taking.)

- Accommodate victims' request for responders of a specific gender as much as possible.

- Prior to starting the exam and before each procedure, describe what is entailed and its purpose to patients. Be sure that communication/language needs are met and information is conveyed in a manner that patients will understand. After providing this information, seek patients' permission to proceed and respect their right to decline any part of the exam. However, follow exam facility and jurisdictional policy regarding minors and adults who are incompetent to give consent. (For a more detailed discussion on seeking informed consent of patients, including consent by victims from specific populations, see *A.3. Informed Consent.*)

- Assess and respect patients' priorities.

- Integrate exam procedures where possible (e.g., blood samples needed for medical and evidentiary purposes should be drawn at the same time).

- Address patients' safety concerns during the exam. Sexual assault patients have legitimate reasons to fear further assaults from their attackers. Local law enforcement may be able to assist facilities in addressing patients' safety needs.

- Provide information that is easy for patients to understand and that can be reviewed at their convenience. (Also see *C.10. Discharge and Followup.*)

- After the exam is finished, provide patients with the opportunity to wash, change clothes (providing clean replacement clothing if necessary), get food or drinks, and make needed phone calls.

3. Informed consent: Patients should understand the full nature of their consent to each exam procedure. By presenting them with relevant information, they are in a position to make an informed decision about whether to accept or decline a procedure. However, they should be aware of the impact of declining a particular procedure, as it may negatively affect the quality of care, the usefulness of evidence collection, and, ultimately, any criminal investigation and/or prosecution. They should understand that declining a particular procedure might also be used to discredit them in court. If a procedure is declined, reasons why should be documented if the patient provides such information.

   Recommendations for healthcare providers and other responders to request patients' consent during the exam process:

   • Seek informed consent as appropriate throughout the exam process for medical evaluation and treatment and the forensic exam and evidence collection. Coordinate efforts to obtain consent among responders.
   • Be aware of statutes and policies governing consent in cases of minor patients, vulnerable adult patients, and patients who are unconscious, intoxicated, or under the influence of drugs. In all cases, however, the exam should never be done against the will of the patient.

4. Confidentiality: Involved responders must be aware of the scope and limitations of confidentiality related to information gathered during the exam process. Confidentiality is intricately linked to the scope of patients' consent. Members of a Sexual Assault Response Team (SART) or other collaborating responders should inform victims of the scope of confidentiality with each responder and be cautious not to exceed the limits of victim consent to share information in each case.

   Recommendations that jurisdictions may take to maintain confidentiality of patients:

   • Make sure that jurisdictional policies address confidentiality related to the medical forensic exam (e.g., of forensic documentation, photographs, and colposcopic video images).
   • Increase responders' and patients' understanding of confidentiality issues (e.g., scope of confidentiality advocates can provide; scope of confidentiality of information shared with examiners, law enforcement, prosecutors, and other responders with whom patient has contact; and what happens to information once it enters the criminal justice system).
   • Consider the impact of Federal privacy laws regarding health information on victims of sexual assault.
   • Strive to resolve intrajurisdictional conflicts.

5. Reporting to law enforcement: Reporting provides the criminal justice system with the opportunity to offer immediate protection to victims, collect evidence from all crime scenes, investigate cases, prosecute if there is sufficient evidence, and hold offenders accountable for crimes

committed. Given the danger that sex offenders pose to the community, reporting can serve as a first step in efforts to stop them from reoffending. Equally important, reporting gives the justice system the chance to help victims address their needs, identify patterns of sexual violence in the jurisdiction, and educate the public about such patterns. It is recommended that service providers encourage victims to report due in part to the recognition that delayed reporting is detrimental to the prosecution and to holding offenders accountable. Victims need to know that even if they are not ready to report at the time of the exam, the best way to preserve their option to report later is to have the exam performed.

Reporting requirements in sexual assault cases vary from one jurisdiction to another. Every effort should be made to facilitate treatment and evidence collection (if the patient agrees), regardless of whether the decision to report has been made at the time of the exam. Victims who are undecided about reporting who receive respectful and appropriate care and advocacy at the time of their exam are more likely to assist law enforcement and prosecution.

Recommendations for jurisdictions and responders to facilitate victim-centered reporting practices:

- Where permitted by law, patients, not healthcare workers, should make the decision to report a sexual assault to law enforcement. Patients should be provided with information about possible benefits and consequences of reporting so that they can make an informed decision.
- It is not recommended to require reporting as a condition of performing or paying for the exam. Even if patients are undecided about reporting, they should be encouraged to provide a medical forensic history, undergo the forensic exam, and have evidence collected and stored.
- Jurisdictions may want to consider alternatives to standard reporting procedures. For example, an anonymous or blind reporting system may be useful in cases in which victims do not want to report immediately or are undecided about reporting.
- Jurisdictions should consider a variety of approaches that promote a victim-centered reporting process.

Payment for the examination under VAWA: Under the Violence Against Women Act (VAWA),* a State, Territory, or the District of Columbia is entitled to funds under the STOP Violence Against Women Formula Grant Program only if it, or another governmental entity, incurs the full out-of-pocket cost of medical forensic exams for victims of sexual assault. The VAWA provisions indicate the exam should minimally include "an examination of physical trauma; determination of penetration/force; a victim interview; and collection and evaluation

* 42 U.S.C. § 3796gg-4.

of evidence."* "Full out-of-pocket costs" means any expense that may be charged to a victim in connection with the exam for the purpose of gathering evidence of a sexual assault.†

Recommendations for jurisdictions to facilitate payment for the sexual assault medical forensic exam:

- Understand the scope of the VAWA provisions related to exam payment.
- Ensure that victims are notified of exam facility and jurisdictional policies regarding payment for medical care and the medical forensic exam, as well as if and how reporting decisions will impact payment. Relevant government entities are strongly encouraged to pay for medical forensic exams regardless of whether victims pursue prosecution.

B. Operational Issues

1. Sexual assault forensic examiners: These are the health care professionals who conduct the examination. It is critical that all examiners, regardless of their discipline, are committed to providing compassionate and quality care for patients disclosing sexual assault, collecting evidence competently, and testifying in court as needed.

   Recommendations for jurisdictions to build the capacity of examiners performing these exams:

   - Encourage the development of specific examiner knowledge, skills, and attitudes.
   - Encourage advanced education and supervised clinical practice of examiners, as well as certification for nurses who are examiners.

2. Facilities: Health care facilities have an obligation to provide services to sexual assault patients. Designated exam facilities or sites served by specially educated and clinically prepared examiners increase the likelihood of a state-of-the-art exam, enhance coordination, encourage quality control, and increase quality of care for patients.

   Recommendations for jurisdictions to build capacity of health care facilities to respond to sexual assault cases:

   - Recognize the obligation of health care facilities to serve sexual assault patients.
   - Ensure that exams are conducted at sites served by specially educated and clinically prepared examiners. A designated facility may employ or have ready access to examiners to conduct the exam. Some

---

* 28 C.F.R. § 90.2(b) (1). The analysis of evidence gathered during the examination, along with examiner documentation of findings, may help in determining whether penetration occurred or force was used. However, examiners are not responsible for drawing conclusions about how injuries were caused or whether the assault occurred or not (although they can note consistency between patients' statements and injuries they identify).
† 28 C.F.R. § 90.14(a).

jurisdictions have examiner programs that serve one or multiple exam sites within a specific area.

- Explore what is best for the community regarding locations for exam sites. It is critical to consider how accessible facilities are to patients disclosing sexual assault, as well as the facility's capacity to properly conduct these exams and treat related injuries.
- Recognize that exam facilities and examiners may benefit from networking with examiners in other facilities or areas for support with peer review of medical forensic reports, quality assurance, and information sharing (e.g., on training opportunities, practices, and referrals for patients).
- Consider developing basic jurisdictional requirements for exam sites.
- Promote public awareness about where exams are conducted. Use specially educated and clinically prepared forensic examiners to conduct the exam, ensuring dissemination of relevant information to appropriate agencies and community members. Encourage first responders to work together to assist victims in using these sites.
- If a transfer from one health care facility to a designated site is necessary, use an established protocol that minimizes time delays and loss of evidence while addressing a patient's needs. However, avoid transferring these patients whenever possible.

3. Equipment and supplies: Certain equipment and supplies are essential to the exam process (although they may not be used in every case). These include a copy of the most current exam protocol used by the jurisdiction, standard exam room equipment and supplies, comfort supplies for patients, sexual assault evidence collection kits, an evidence drying device/method, a camera, testing and treatment supplies, an alternate light source, an anoscope, and written materials for patients. A microscope and/or toluidine blue dye may be required, depending on jurisdictional policy. A colposcope or other magnifying instrument is strongly suggested. Some jurisdictions are also beginning to use advanced technology (telemedicine), which allows examiners offsite consultation with medical experts by using computers, software programs, and the Internet.

Recommendations for jurisdictions and responders to ensure that proper equipment and supplies are available for examinations:
- Consider what equipment and supplies are essential.
- Address cost barriers to obtaining equipment and supplies.

4. Sexual assault evidence collection kit (for evidence from victims): Most jurisdictions have developed their own sexual assault evidence collection kits or purchased premade kits through commercial vendors. Kits often vary from one jurisdiction to another. Despite variations, however, it is critical that every kit meets or exceeds minimum guidelines for contents: broadly including a kit container, instruction sheet and/or checklist, forms, and materials for collecting and preserving all evidence required by the

applicable crime laboratory. Evidence that may be collected includes, but is not limited to, clothing, foreign materials on the body, hair (including head and pubic hair samples and combings), oral and anogenital swabs and smears, body swabs, and a blood or saliva sample for DNA analysis and comparison. The instruction sheet and/or checklist should guide examiners on maintaining the chain of custody for evidence collected.

Recommendations for jurisdictions and responders when developing/customizing kits:

- Use standardized kits (across a local jurisdiction, region, State, Territory, or tribal land) that meet or exceed minimum guidelines for contents, as described above.
- Make kits readily available at any facility that conducts sexual assault medical forensic exams.
- Periodically review the kit's efficiency and usefulness and make changes as needed.

5. Timing considerations for collecting evidence: Although many jurisdictions currently use 72 hours after the assault as the standard cutoff time for collecting evidence, evidence collection beyond that point is conceivable. Because of this, some jurisdictions have extended the standard cutoff time (e.g., to 5 days or 1 week). Advancing DNA technologies continue to extend time limits because of the stability of DNA and sensitivity of testing. These technologies are even enabling forensic scientists to analyze evidence that was previously unusable when it was collected years ago. Thus, it is critical that in every case where patients are willing, examiners obtain the medical forensic history, examine patients, and document findings. Not only can the information gained from the history and exam help health care providers address patients' medical needs, but it can guide examiners in determining whether there is evidence to collect and, if so, what to collect.

Recommendations for health care providers and other responders to maximize evidence collection:

- Whether or not evidence is collected, examiners should obtain the medical forensic history as appropriate, examine patients, and document findings (with patients' consent). Patients' demeanor and statements related to the assault should also be documented.
- Promptly examine patients to minimize loss of evidence and to identify medical needs and concerns.
- Decide whether to collect evidence and what to collect on a case-by-case basis, remembering that outside time limits for obtaining evidence vary.
- In any case, where the need for evidence collection is in question, encourage dialogue about the potential benefits or limitations of collection. Avoid basing decisions about whether to collect evidence on a patient's characteristics or circumstances (e.g., the patient has used illegal drugs).

- Responders should seek education and resources that aid them in making well-informed decisions about evidence collection.
6. Evidence integrity: Properly collecting, preserving, and maintaining the chain of custody of evidence is critical to its subsequent use in criminal justice proceedings.

    Recommendations for health care providers and other responders to maintain evidence integrity:
    - Follow jurisdictional policies for drying, packaging, labeling, and sealing the evidence.
    - Follow jurisdictional policies for documenting exam findings, the medical forensic history, and the patient's demeanor/statements, and packaging, labeling, and sealing such documentation.
    - Follow jurisdictional policies for consistent evidence management and distribution. A duly authorized agent should transfer evidence from the exam site to the appropriate crime lab or other designated storage site (e.g., a law enforcement property facility).
    - Make sure storage procedures maximize evidence preservation. Ensure that storage areas are kept secure and at the proper temperature for the evidence. Also, make sure jurisdictional policies are in place to address the secure storage of evidence in cases in which patients are undecided about reporting.
    - Maintain the chain of custody of evidence. All those involved in handling, documenting, transferring, and storing evidence should be educated regarding the specifics of their roles in properly preserving evidence and maintaining the chain of custody.

C. The Examination Process

1. Initial contact: Some sexual assault patients may initially present at a designated exam facility, but most who receive immediate medical care initially contact a law enforcement or advocacy agency for help. If 911 is called, law enforcement or emergency medical services (EMS) may be the first to provide assistance to victims. Communities need to have procedures in place to promptly respond to disclosures/reports of sexual assault in a standardized and victim-centered manner.

    Recommendations for jurisdictions and responders to facilitate initial contact with victims:
    - Build consensus among involved agencies regarding procedures for a coordinated initial response when a recent sexual assault is disclosed or reported, and educate responders on procedures. Encourage victims to interact with advocates as soon after disclosure as possible.
    - Recognize essential elements of initial response. In particular, encourage victims to seek medical care and have evidence collected. In the case of life-threatening or serious injuries, obtain emergency medical assistance according to jurisdictional policy. Any life-threatening

wounds should be treated and victims' immediate safety needs should be addressed before evidence is collected.

- If victims decide to seek medical care and/or have evidence collected, follow jurisdictional policies for preserving evidence, collecting a urine sample if needed, and transporting victims to the exam site.

2. Triage and intake: Once patients arrive at the exam site, health care personnel must evaluate, stabilize, and treat for life-threatening and serious injuries according to facility policy. Standardized procedures for response in these cases should be followed, while respecting patients and maximizing evidence preservation.

Recommendations for health care providers to facilitate triage and intake that addresses patients' needs:

- Consider sexual assault patients a priority. Use private locations in the exam facility for the primary patient consultation and initial law enforcement interviews, offer a waiting area for family members and friends, and provide childcare if possible.
- Respond to acute injury, trauma care, and safety needs of patients before collecting evidence. Patients should not wash, change clothes, urinate, defecate, smoke, drink, or eat until initially evaluated by examiners, unless necessary for treating acute medical needs.
- Alert examiners to the need for their services at the exam site.
- Contact victim advocates so they can offer services to the patient, if not already done.
- Assess and respond to safety concerns, such as threats to the patient or staff, upon arrival of patients at the exam site.
- Assess patients' needs for immediate medical or mental health intervention. Seek informed consent from patients before providing treatment, according to facility policy.

3. Documentation by health care personnel: Examiners document exam findings, the medical forensic history, and evidence collected in the medical forensic report. Examiners and/or other involved clinicians separately document medical care in the patient's medical record.

Recommendations for health care providers to complete needed documentation:

- Ensure completion of all appropriate documentation. The forensic details of the exam are documented in the medical forensic report, according to jurisdictional policy. The only medical issues documented in this report are acute findings that potentially relate to the assault or preexisting medical factors that could influence interpretation of findings. Separate medical documentation by examiners and other involved clinicians follows a standard approach—address acute complaints, gather pertinent historical data, describe findings, and document treatment and followup care.
- Ensure the accuracy and objectivity of medical forensic reports by seeking education on proper report writing.

4. The medical forensic history: Examiners ask the patient questions to obtain this history. This information guides them in examining the patient and collecting evidence.

Recommendations for health care providers to facilitate gathering information from patients:

- Examiners should coordinate with other responders, primarily law enforcement representatives, to facilitate information gathering that is respectful to patients and minimizes repetition of questions.
- Keep in mind that advocates may support and advocate for patients when the medical forensic history is taken (if desired by patients), but they may not actively participate in the process. Patients should be informed that the presence of family members, friends, and others offering personal support during this time may influence or be perceived as influencing their statements. If patients choose to have others present despite this knowledge, these individuals should not actively participate in the process.
- Consider and address patients' needs prior to information gathering, including identifying the level of their communication skill and modalities and then tailoring information gathering accordingly.
- Obtain the medical forensic history in a private, quiet setting.
- Gather information for the history according to jurisdictional policy. Include the date and time of the assault, pertinent patient medical history (e.g., menstruation history), recent consensual sexual activity of the patient, the patient's activities since the assault (e.g., took a shower), the patient's assault-related history (e.g., loss of consciousness), suspect information, if known (e.g., number and gender of assailants), nature of the physical assault, and description of the sexual assault.

5. Photography: Photographic evidence of injury on the patient's body can supplement the medical forensic history and document physical findings.

Recommendations for health care providers and other responders to photograph evidence:

- Come to a consensus about the extent of forensic photography necessary. Some jurisdictions routinely take photographs of both detected injuries on patients and normal (apparently uninjured) anatomy, while others limit photography to detected injuries.
- Consider who will take photographs and what equipment will be used. Photographers should be familiar with equipment operation as well as educated in forensic photography and in ways to maintain the patient's privacy and dignity while taking photographs. Consult with jurisdictional criminal justice agencies and examiners regarding the type of equipment that should be used.
- Obtain informed consent from patients before taking photographs. Patients should understand the purpose of the photographs, what will be photographed and any related procedures, the potential uses

of photographs during investigation and prosecution, and the possible need for followup photographs.

- Consider the patient's comfort and need for modesty.
- Identify who will be present when photographs are taken.
- Take initial and followup photographs as appropriate, according to jurisdictional policy.

6. Exam and evidence collection procedures: Examiners examine patients and collect evidence according to jurisdictional policy. Findings from the exam and collected evidence often help reconstruct the events in question in a scientific and objective manner.

Recommendations for health care providers to conduct the exam and facilitate evidence collection:

- Strive to collect as much evidence from patients as possible, considering the scope of informed consent, the medical forensic history, the examination, and evidence collection kit instructions.
- Be aware of evidence that may be pertinent to the issue of whether the patient consented to sexual contact with the suspect. Understand how biological evidence is tested.
- Prevent exposure (of both patients and staff) to infectious materials and contamination of evidence.
- Understand the implication of the presence or lack of semen (in cases involving male suspects).
- Seek informed consent from patients for each portion of the exam and evidence collection.
- Modify the exam and evidence collection to address the specific needs and concerns of patients.
- Conduct the general physical and anogenital examination, guided by the scope of informed consent and the medical forensic history. Document findings on body diagram forms. With the patient's consent, use an alternate light source, colposcope, and anoscope, as appropriate and if available, to increase the likelihood of detecting evidence.
- Collect evidence to submit to the crime lab for analysis, according to jurisdictional policy.
- Collect blood and/or urine for toxicology screening, if applicable.
- Keep medical specimens separate from forensic specimens collected during the exam.

7. Drug-facilitated sexual assault: Responders must consider the possibility that drugs may have been used to facilitate an assault. They must know how to screen for suspected drug-facilitated sexual assault, obtain informed consent of patients for testing, and collect toxicology samples when needed.

Recommendations for jurisdictions and responders to facilitate response in suspected drug-facilitated sexual assault:

- Educate examiners, 911 dispatchers, law enforcement representatives, prosecutors, judges, and advocates on related issues. Develop jurisdictional policies to clarify first responders' roles in cases involving suspected drug-facilitated assault.

- Be clear about the circumstances in which toxicology testing may be indicated (for optimal care or when there is a suspicion of drug-facilitated sexual assault). Routine toxicology testing in all sexual assault cases is not recommended.
- Informed consent of patients should be sought to collect toxicology samples. Patients should be aware of the purposes and scope of testing that will be done, potential benefits and consequences of testing, any followup treatment necessary, how they can obtain results, who will pay for the testing, and if they have any opportunity to revoke consent to testing.
- With patients' permission, immediately collect a urine specimen if it is suspected that ingestion of drugs used to facilitate sexual assault occurred within 96 hours prior to the exam. The first available urine should be collected—law enforcement and emergency medical services should be trained and prepared to collect a urine sample if patients must urinate prior to arrival at the health care facility for the exam. Advocates and other professionals who may have contact with patients prior to their arrival at the exam site should also be educated to provide those who suspect drug-facilitated assault with information on how to collect a sample if the patient cannot wait to urinate until getting to the site.
- Also, collect a blood sample if it is suspected that the ingestion of drugs used to facilitate sexual assault occurred within 24 hours of the exam. If a blood alcohol determination is needed, collect blood within 24 hours of ingestion of alcohol, according to jurisdictional policy.
- Jurisdictional policies should be in place and followed for packaging, storing, and transferring samples.

8. Sexually transmitted infection (STI) evaluation and care: Because contracting an STI from an assailant is of significant concern to patients, it should be addressed during the exam.

   Recommendations for health care providers to facilitate STI evaluation and care:

   - Offer patients information about the risks of STIs (including HIV), the symptoms and what to do if symptoms occur, testing and treatment options, followup care, and referrals. Referrals should include free and low-cost testing, counseling, and treatment available in various sections of the community. For HIV testing, confidential and anonymous testing is recommended.
   - Consider testing patients for STIs during the initial exam on a case-by-case basis. If testing is done, follow the guidelines of the Centers for Disease Control and Prevention (CDC).
   - Encourage patients to accept prophylaxis against STIs during the initial exam. (Note, however, that treatment may not be appropriate

for some individuals—for example, if they have a condition that may be adversely affected by taking prophylaxis.) The CDC suggests a regimen to protect against chlamydia, gonorrhea, trichomonas, and bacterial vaginosis (BV), as well as the hepatitis B virus. If accepted, provide care that meets or exceeds CDC guidelines. If declined, it is medically prudent to obtain cultures and arrange for a followup exam and testing. Seek informed consent from patients for treatment, according to facility policy.

- Encourage and facilitate followup STI examinations, testing, immunizations, and treatment as directed.
- Offer postexposure prophylaxis for HIV to patients at high risk for exposure, particularly when it is known that suspects have HIV/AIDS. Meet or exceed CDC recommendations. Discuss risks and benefits of the prophylaxis with patients prior to their decisions to accept or decline treatment. Careful monitoring and followup by a health care provider or agency experienced in HIV issues is required.

9. Pregnancy risk evaluation and care: Female patients may fear becoming pregnant as a result of an assault. Health care providers must address this issue according to facility and jurisdictional policy.

Recommendations for health care providers to facilitate pregnancy evaluation and care:

- Discuss the probability of pregnancy with patients.
- Administer a baseline pregnancy test for all patients with reproductive capability.
- Discuss treatment options with patients, including reproductive health services.

10. Discharge and followup: Health care personnel have specific tasks to accomplish before discharging patients, as do advocates and law enforcement representatives (if involved). Responders should coordinate discharge and followup activities as much as possible to reduce repetition and avoid overwhelming patients.

Recommendations to facilitate discharge and followup:

- It is important to ensure that patients are fully informed about postexam care. Information may include referrals to other professionals to make sure that patients' medical and/or mental health needs related to the assault have been addressed, discharge instructions, followup appointments with the examiner or other health care providers, and contact procedures for medical followup. In addition to medical followup, followup may be indicated to document developing or healing injuries and complete resolution of healing.
- Advocates and law enforcement representatives, if involved, should coordinate with examiners to discuss other issues with patients, including planning for their safety and well-being, physical comfort needs, information needs, the investigative process, advocacy and

counseling options, and law enforcement and advocacy followup contact procedures.

11. Examiner court appearances: Health care providers conducting the exam should expect to be called on to testify in court as fact and/or expert witnesses.

Recommendations for jurisdictions to maximize the usefulness of examiner testimony in court:

- Encourage broad education for examiners on testifying in court.
- Promote prompt notification of examiners if there is a need for them to testify in court.
- Encourage pretrial preparation of examiners.
- Encourage examiners to seek feedback on testimony to improve effectiveness of future court appearances (USDOJ, A National Protocol [SAFEta] 2004).

Appendix 5.1: Suggestions for Police Officers in Sexual Assault Cases

Law enforcement exemplifies the concept of public service. The motto is to serve and protect. Those goals are achieved in a variety of ways, including physically protecting a person or their property, interviewing victims and witnesses, collecting and securing evidence, analyzing evidence and maintaining it for presentation in court, testifying, and helping prevent crime and violence in our communities.

In cases of sexual assault, the long-term psychological consequences usually outweigh the physical ones. Therefore, focusing on the victim before and during your handling of the case is crucial to the long-term outcome.

Although the following comments may seem like common sense to the experienced officer, other officers may be relatively unfamiliar and/or uncomfortable in dealing with victims of sexual violence. These guidelines are intended to provide all officers with a reminder of how to best handle the victims of sexual crime.

Many victims are quite unfamiliar with the victim experience and our criminal justice system. Initially it all appears to be quite impersonal, frightening, and overwhelming. While a successful prosecution will go a long way to help a survivor move toward a positive resolution and may help to reduce the overall incidence of sex crimes in our communities, if the victim is inadvertently intimidated by the system, she is less likely to cooperate and follow through with prosecution.

1. Be mindful of the severe psychological trauma the victim has just suffered. Approach the victim in an unhurried, professional manner; don't "charge in" like a rescuing cowboy. A gentle, supportive demeanor works best in these cases.
2. Speak softly and avoid any aggressive or intimidating behavior that might further frighten the victim.
3. If you are first on-scene make sure EMS is in route and let the victim know they should be arriving shortly. If EMS is not coming, accompany the victim to medical care. Explain basic procedures to help alleviate the victim's fears.
4. Let the victim know you are concerned for her safety and well-being. Try to determine whether the perpetrator is still nearby or might return soon. In addition to time and place of assault, try to obtain information on the perpetrator to assist in apprehension: physical description, clothing worn, vehicle used (if any), direction of flight and any weapon used. Transmit this basic information by radio, if possible, but do not release the victim's name or address. Local media DO listen to radio transmissions.

5. If the assault was recent, advise the victim not to wash, douche, or urinate. Explain that valuable evidence may otherwise be lost.

6. Advise the victim that sexual assault nurse examiners (SANE nurses) are available and that they specialize in sexual assault cases, including confidential interviews, evidence collection, pregnancy and sexually transmitted disease prophylaxis, and follow-up review.

7. Photographs and evidence collection may be done at the scene, but examination of the victim's body and wound photographs will be taken by the SANE nurse. A national protocol is available free online from SAFEta (http://www.safeta.org/index.cfm). Explain procedures in your jurisdiction and include a mention of this in your notes.

8. Avoid in-depth questioning until meeting up with the sexual assault nurse examiner. If you will not be assigned to this case, leave such questioning to that officer. This reduces the number of times the victim has to initially tell her story.

9. Throughout contact with the victim and/or family, avoid the appearance of being judgmental and making statements that might be construed as prejudicial. Regardless of any poor choices made by the victim, including the choice to be in a vulnerable situation, now is not the time to berate or "educate" the victim.

10. Remember, the family has also been affected by this event. Treat the family with the same respect and caution as the victim.

11. Try to conduct interviews in a room or space different from the actual "scene" of the attack. This will help to reduce inadvertent contamination of the scene.

12. Interview family and others present separately. Be sure to let them know that the victim is blameless and acted correctly in submitting. Otherwise, she might have been killed.

13. Protect the victim's anonymity by minimizing exposure to unwanted attention from bystanders, media, or others.

14. If the victim presents to the police station, determine if a family member may have perpetrated this crime and contact the local victim's advocacy center or other community support agency to find the victim a safe place to stay.

16. If the victim specifically and continuously requests a female officer, make every effort to provide one. There are, however, advantages to having a sensitive male deal with sex crimes. In addition to showing the victim that not all men are aggressive and violent, a supportive male can connect with a female victim in a way that another female cannot. Additionally, a supportive officer can ease the transition between fear of males and trust. The gender of the investigating officer is not nearly as important as their attitude, approach, sensitivity, and skills in crisis intervention and investigative competence.

17. Especially for victims who choose not to have a medical evaluation and consult, initiating contact with victim support services such as the local rape crisis center is important in the victim's recovery. Inform the victim of locally available counseling and support services and recommend they take advantage of these services. Many of them are free.

18. Do not try to explain investigative or judicial procedures at this time. When in crisis, victims remember little of what happened and almost none of what was said. In addition, the victim will need support and guidance throughout the entire process. Counseling services, victim support providers, and legal advocacy services are designed for these purposes.

19. Remember, the immediate safety, support, and treatment are of paramount concern.

Key Point:
The actions of the first officer on scene may have a vital impact on the future psychological well-being of the victim and may play a significant role in deciding whether or not to follow through with filing charges and prosecution.

Victims of sexual assault will never be the same. Memory of the assault will fade with time and support, but it will never be erased. Life will go on. Hopefully, the rape will become something that happened but not define the victim's entire identity. By recognizing the need for appropriate sensitivity, all providers assist victims by helping them make the transition from victim to survivor.

Source: Adapted. Original information contained in North Coast Forensics, Richfield, Ohio, FBI training manual, 1992. Additions and suggestions from Sgt. George Kral, TPD, Personal Assault Unit, John W. Calogar, retired Chief Olmsted Falls, Ohio Fire Dept., Sally Royston, RN, SANE Coordinator, Toledo Area SANE/SART Program. Secondary source unknown.

Appendix 5.2: Suggestions for First Responders in Sexual Assault Cases

Fire, rescue, and emergency medical service providers are the people we turn to in a time of crisis. Like law enforcement, these public servants provide the unseen, and hopefully unneeded, security net supporting our health and daily activities. The goals of emergency services are achieved in a variety of ways, from rescuing a cat or stranded skier to getting the heart attack victim to definitive care at a hospital, to caring for the victims of trauma, interpersonal violence, or sexual assault, and from preventing fires to putting them out. Be it in the home, at work, on the street or out in the wilderness, fire, rescue, and EMS personnel are the knights in shining armor in times of crisis.

In cases of sexual assault, the long-term psychological consequences usually outweigh the physical ones. Therefore, focusing on the victim's psychological needs is as important as your handling of the case to long-term health and welfare of the patient.

Although the following comments may seem like common sense to the experienced provider, other healthcare providers may be relatively unfamiliar and/or uncomfortable in dealing with victims of sexual violence. These guidelines are intended to provide all providers with a reminder of how to best handle the victims of sexual crime.

1. Many victims are quite unfamiliar with the victim experience and our criminal justice system. It initially appears to be quite impersonal, frightening, and overwhelming. While a successful prosecution will go a long way to help a survivor move toward a positive resolution and may help to reduce the overall incidence of sex crimes in our communities, if the victim is inadvertently intimidated by the system she is less likely to follow through to see prosecution finalized.
2. Be mindful of the severe psychological trauma the victim has just suffered. Approach the victim in an unhurried, professional manner; don't charge in like a rescuing cowboy. A gentle, supportive demeanor works best in these cases.
3. Speak softly and avoid any aggressive or intimidating behavior that might further frighten the victim.
4. Let the victim know you are concerned for her safety and well-being and try to determine whether the perpetrator is still nearby or might return soon. If there is any chance the perpetrator might return or the scene is unsafe in any other way (e.g., there is a gas leak or other danger), call law enforcement immediately and remove yourself, your crew, and the victim to a safe location.
5. Do your primary assessment.

6. If life-threatening conditions exist, load and go.
   a. Please try to avoid cutting through stains, cuts, or other places where evidence might be found.
   b. Have dispatch advise the receiving facility that you have a sexual assault victim.
   c. If the assault was recent, advise the victim not to wash, douche, or urinate. Explain that valuable evidence may otherwise be lost.
   d. Have dispatch advise the local SANE unit or healthcare facility that a sexual assault victim is on the way.
7. If life-threatening conditions do not exist, finish your assessment and determine if the patient needs transport to the hospital or SANE unit and if she wishes to be transported by rescue or wishes to go on her own.
   a. Recognize that the patient's body may be a crime scene containing evidence.
   b. If the assault was recent, advise the victim not to wash, douche, urinate, or change clothes. Explain that valuable evidence may otherwise be lost. Notify the SANE unit or hospital if the patient has urgent need to void.
   c. Do not use alcohol or germicidal agents, especially in areas where there are signs of biomaterial or where the patient indicates the perpetrator may have left body fluids by licking, spitting, ejaculating, etc. Pick a different spot to start the IV.
   d. If the patient has not changed clothes, leave them on. Please do not cut clothing where evidence might be found—underwear, through tears, rips or cuts, and any area where stains might be found.
   e. Unless there is a reason* not to wait for law enforcement, let them collect evidence. If the patient has already changed clothes, wearing appropriate PPE, collect each item separately and place it in a paper bag. If paper containers are not readily available, notify law enforcement of the whereabouts of all clothing (including shoes, underwear, etc.). Evidence rapidly decomposes in plastic bags and the chain of custody is all-important, so law enforcement should be involved and evidence properly collected.
8. Advise the victim that sexual assault nurse examiners (SANE nurses) are available and that they specialize in sexual assault cases, including confidential interviews, evidence collection, pregnancy and sexually transmitted disease prophylaxis, and follow-up review.

---

* Including length of time for them to arrive and/or demonstrated lack of competence in evidence collection.

9. Protect the victim's anonymity by minimizing exposure to unwanted attention from bystanders, media, or others.

10. Appropriate photographs and evidence collection may be done at the scene, but examination of the victim's body and wound photographs will be taken by the SANE nurse. A body diagram and good narrative should suffice if the patient is going to a medical facility.

11. Explain procedures in your jurisdiction and include a mention of procedures followed in your notes. (A National Protocol, USDOJ, SAFEta, 2004.)

12. Obtain a basic recitation of the assault, but avoid in-depth questioning as that will be done by law enforcement and the sexual assault nurse examiner. This reduces the number of times the victim has to re-live the experience.

13. Throughout contact with the victim and/or family, avoid the appearance of being judgmental and making statements that might be construed as prejudicial. Regardless of any poor choices made by the victim, including the choice to be in a vulnerable situation, now is not the time to berate or "educate" the victim.

14. Remember, the family has also been affected by this event and may be in a state of crisis themselves, especially if the victim is young. Treat the family with the same respect and caution as the victim. If appropriate, let the family know that the victim acted correctly as she is still alive.

15. Try to conduct the patient interview in a room or space different from the actual "scene" of the attack. This will help to reduce inadvertent contamination of the scene.

16. Interview family and others present separately. Be sure to let them know that the victim is blameless and acted correctly in submitting. Otherwise, she might have been killed.

17. If the victim specifically and continuously requests a female provider, make every effort to provide one. There are, however, advantages to having a sensitive male deal with sex crimes. In addition to showing the victim that not all men are aggressive and violent, a supportive male can connect with a female victim in a way that another female cannot. Additionally, a supportive male provider can ease the transition between fear of males and trust. The gender of the healthcare provider is not nearly as important as their attitude, approach, sensitivity, and skills in crisis intervention and investigative competence.

18. Especially for victims who choose not to have a medical evaluation and consult, initiating contact with victim support services such as the local rape crisis center is important. Have a list of local counseling and support services and recommend the victim contact them,

especially if they are not going to a SANE unit or other medical care facility.

19. Do not try to explain investigative or judicial procedures at this time. When in crisis, victims remember little of what happened and almost none of what was said. In addition, the victim will need support and guidance throughout the entire process. Counseling services, victim support providers, and legal advocacy services are designed for these purposes.

20. Remember, the immediate safety, support, and treatment are of paramount concern.

Key Point:
The actions of the first responder on scene may have a vital impact on the future psychological well-being of the victim and may play a significant role in deciding whether or not to follow through with pressing charges and following up with prosecution.

Victims of sexual assault will never be the same. Memory of the assault will fade with time and support but it will never be erased. Life will go on. Hopefully, the rape will become something that happened but not define the victim's entire identity. By recognizing the need for appropriate sensitivity, all providers assist victims by helping them make the transition from victim to survivor.

Adapted from: Original information contained in North Coast Forensics, Richfield, Ohio, FBI training manual, 1992. Additions and suggestions from Sgt. George Kral, TPD, Personal Assault Unit, John W. Calogar, retired Chief Olmsted Falls, Ohio Fire Dept., Sally Royston, RN, SANE Coordinator, Toledo Area SANE/SART Program. Secondary source unknown.

References

Kilpatrick, D. G., C. N. Edmunds, and A. K. Seymour. 1992. *Rape in America: A report to the nation*. National Victim Center and Crime Victims Research and Treatment Center. April 23, 1992.

National Victim Assistance Academy Manual (NVAA). 2002. http://www.ovc.gov/assist/nvaa2002/ (accessed August 28, 2010).

Sexual Assault: Frequently asked questions. http://www.womenshealth.gov/faq/sexualassault.pdf (accessed October 18, 2008).

Tjaden, P. and Thoennes, N. 1998. *The National Women's Study: Research in Brief*. Office on Women's Health, Department of Health and Human Services. Content last updated March 1, 2009.

U. S. Code, Title 18, Chapter 109A, Sexual Abuse, §2241-2243 http://www.law.cornell.edu/uscode/uscode18 (accessed September 2, 2010).

U.S. Department of Justice (USDOJ), Office of Violence against Women. A National Protocol for Sexual Assault Medical Forensic Examinations for Adults and Adolescents. President's DNA Initiative, NCJ 206554. 42 U.S.C. §3796gg-4(d) September 2004. http://www.safeta.org/index.cfm and from http://www.safeta.org/associations/8563/files/National%20Protocol.pdf (accessed August 31, 2010), and http://www.ncjrs.gov/pdffiles1/ovw/206554.pdf (accessed September 2, 2010). Also available free of charge from SAFEta as a CD.

Women's Health. 2010. The National Women's Health Information Center, U.S. Department of Health and Human Services, Office on Women's Health. http://www.womenshealth.gov/faq/support-assault.pdf (accessed August 28, 2010).

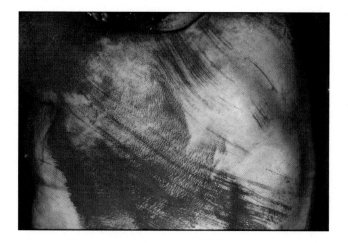

Figure 2.6 Grazes. Courtesy Dr. Patrick Besant-Matthews. Used with permission.

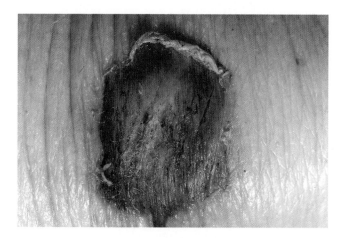

Figure 2.7 Abrasion with accumulated tissue at one end showing direction of force. Courtesy Dr. Patrick Besant-Matthews. Used with permission.

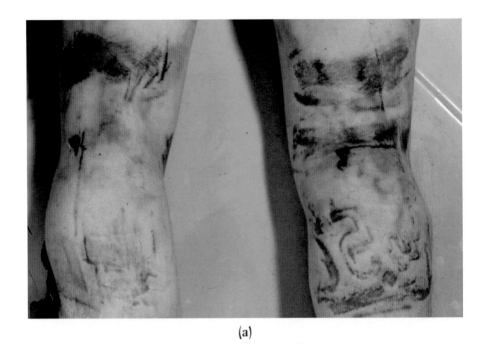

(a)

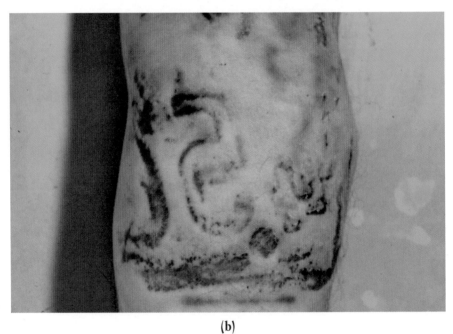

(b)

Figure 2.13a,b Pattern injury. Imprint of license plate on victim's legs where struck by vehicle. Courtesy of Dr. Patrick Besant-Matthews. Used with permission.